Basic Systems Analysis

Basic Systems Analysis

Third Edition

Edited by

Alan Daniels

Former Director
Institute for Industrial Training
Brunel University

Don Yeates

Director of Personnel, Training and Development
The CAP Group plc

Pitman

PITMAN PUBLISHING
128 Long Acre, London WC2E 9AN

© Alan Daniels and Don Yeates, 1982, 1983, 1984, 1988

First published in Great Britain 1982
Reprinted 1982, 1983
Second edition 1984
Reprinted 1985, 1986, 1987
Third edition 1988

British Library Cataloguing in Publication Data

Daniels, Alan
  Basic systems analysis. — 3rd ed.
  1. Systems analysis
  I. Title   II. Yeates, Don
  003   QA402

ISBN 0-273-02931-2

Printed and bound in Great Britain

Contents

Preface and acknowledgements

This third edition includes further improvements that reflect the changes being made to the way systems analysts work. By far the most significant changes are the inclusion of an additional chapter showing a case study using SSADM, and a much increased chapter on structured analysis and design methods. This material therefore now describes all of the stages in SSADM with examples.

The chapter on micros has been completely rewritten and now includes local area networks as well. We've also taken the opportunity to change much of the technology based material and to put some new ideas into chapter one. As usual, we have been helped by friends and colleagues in Datasolve's Education division, especially Don Wilkinson, Alan Peck and Mike Brough. Our word processing expert this time was Susan Punter. The SSADM forms were drawn by Susan Swanson, and the index was prepared by Jan Yeates.

Alan Daniels
Don Yeates

Introduction

The objective of the material in this book is to act as back-up textbook material to all basic courses on systems analysis and in particular to courses on systems analysis leading to the NCC's Certificate in Systems Analysis.

There have been rapid technological developments in computing over the past decade and this book takes into account such aspects as sophisticated methods of data input and output, the advent of the microchip and the many new computer applications now feasible with the introduction of cheaper computing hardware.

Basically the material provides a background for people working as computer programmers or in business disciplines and so enables them to enter the field of business systems analysis. It is also relevant for experienced business people and managers wishing to understand the use of computers in business and the problems of introducing computer-based systems into business and administration.

From Chapter 1, the book sets out to define the scope of the systems analyst and the problems encountered in introducing a computer or a new computing system into an organization where people are suspicious of, or even antagonistic to, the introduction of new technology.

Chapter 2 deals with systems investigation and analysis, i.e. the work of the analyst in fact-finding, fact-recording and the ability to describe in documentary form the essential operational requirements of a business system.

Chapters 3 and 4 consider the problems of output and input design.

Chapter 5 is concerned with file design, various types of storage device and various methods of file organization.

Chapter 6 deals with the traditional approach to systems design. It examines common problems of design and then discusses the establishment of a standard method of documenting the agreed design.

Chapter 7 deals with aspects of controls and security in the system with special attention given to auditing.

Chapter 8 deals with methods of proving the finished design, testing and timing.

Chapter 9 looks at the problem of system justification, i.e. the computer justifies itself by achieving the desired results taking into account the system objectives and feedback to ensure that these objectives are achieved.

Chapter 10 emphasizes that throughout the whole process, communication and clear report writing are essential aspects of systems implementation.

Chapter 11 deals with the implementation itself covering the ground from the acceptance of the design to its satisfactory operation supported by appropriate user and operation manuals.

Chapters 12 and 13 deal with general aspects of hardware, software, and data communications. The hardware is dealt with in broad outline only since hardware requirements will vary and the manufacturer will supply detailed technical literature on the capabilities of various pieces of equipment. New material on storage and input/output has been included.

Chapter 14 is an attempt to put the 'microprocessor revolution' into context and to remove many of the myths associated with micros with a brief look into the future to see how such systems are likely to evolve. This is a completely revised chapter in this edition.

Chapter 15 is an introduction to the newer structured methods for analysis and design which now play an important part in the way systems work is done. Detailed information is included about SSADM.

Chapter 16 is a worked case study example using SSADM.

Chapter 17 deals with the basics of accounting principles for financial and management accounting.

The book assumes no previous knowledge of systems analysis technique and is suitable for both students and programmers wishing to pursue a course in systems analysis, or for business people wishing to acquaint themselves with the basic problems of introducing the computer into an organization. Managers should particularly note Chapters 1, 3, 4, 7, 9, 10 and 14.

Alan Daniels
Don Yeates

1 The Scope of Systems Analysis

Before one can study the techniques a systems analyst must understand and apply, one must appreciate the systems analyst's function in the wider context of the employing organization. Regardless of what purpose the organization serves, a systems analyst must have certain fundamental personal qualities to succeed.

This book is directed primarily toward new entrants to the field of systems analysis. The following brief dicussion of the analyst's purpose is intended to establish a basis of agreement about what analysts seek to be and do. This definition of purpose will help the new analyst to see the total activity in perspective before becoming immersed in the finer details of the procedures which are described in subsequent chapters.

1.1 Organizational background

Systems within an organization do not exist in a vacuum. They reflect the organization's structure and purpose, and to a lesser degree are influenced by the personalities of both management and staff concerned. When the systems analyst is to start work on a particular project, an informed awareness of the employer's organization is necessary. This will help during employee interviews and will also assist in placing details of the current system in perspective. Often insufficient time is devoted to this area of training. As a result, the analyst does not comprehend fully the significance of some matters discovered during the fact-finding phase of the work, and may design the new system with the limitations of the old.

There must also be an understanding of the constitution of the organization. This covers the overall arrangement of its constituent parts, and the strength of its affiliations with other organizations. When the organization involved is diversified across more than one location, the analyst should discover the function of each.

The analyst will benefit from a knowledge of the organization's history. By tracing the major milestones in the evolution of the firm, the type of management decisions that were made in the past can be identified. The management's overall policies should be kept in mind. The company's annual reports, employee handbooks, the trade literature, investment reviews and press clippings should be examined. This background will help the analyst to appreciate the precise divisional structures currently existing at each location. From this it can be discovered how much of the structure is carefully planned and how much is dependent on former structures. At the same time an attempt should be made to discover how closely the present objectives of each main function are in accord with the policies now being followed.

As part of the fact-finding operations, the analyst will need to detail the departmental structure of the area in which the study is being carried out. If there is a company organization chart, the analyst should ascertain that it reflects current conditions and become acquainted with it. If no chart exists, then one should be prepared. In any case, some clear distinction should be drawn between those operations that involve line management and those that are staff functions. The former are responsible for executing policies and translating them into attainable objectives, while the latter usually act as advisors.

Conventional organization charts, which show a number of branches from a chief executive or board of directors at the top, represent the pyramid of control found in most organizations. The analyst should superimpose on this structure another one that describes the true lines of communication. This structure may not follow the strict divisional and departmental boundaries implied by the original chart, but it will show the analyst how the proposals introduced in one area may affect the work in others. Probably much of the material mentioned above is recorded in one form or another, and will have to be edited and distilled to obtain a coherent picture of the organization. However, there will probably not be any recorded information about the environment in which the organization operates. This information is needed because the firm itself has little direct influence on its environment, which is constantly changing.

Some examples of facts the systems analyst should try to discover are the organization's position within its industry, the labour relationships in the industry, and the effect of government policies on the industry. A new system may operate no better than a previous system, although it may appear much better on paper, simply because some environmental condition frustrates its operation. The systems analyst may find it necessary to estimate and predict environmental conditions to anticipate flexibility of the system designed.

1.2 Applications development and planning

The development of new computer based applications is intended to enable organisations to operate more efficiently. Computer based systems process data to produce information so that better decisions can be taken. Data needed for decision making is collected, filtered and summarised upwards. The sources and flow of information are shown in Figure 1.2.1.

The development of a new application begins with a user request described formally in a Statement of Requirements. Development of a new system passes through various stages until a working system is produced. We now distinguish between 'traditional' approaches to systems development and more recent 'structured' approaches. The traditional approach is described below with the various stages, their activities and the end products of each stage.

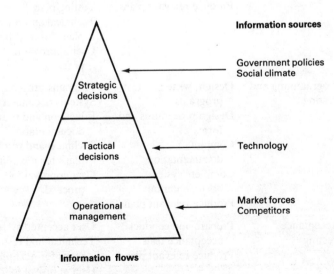

Fig. 1.2.1   Sources and flow of information

4 Basic Systems Analysis

Stage	Activities	End Products
1 Business Survey	Define problem Investigate user requirements Recommend course of action	Preliminary report
2 Feasibility Study	Prepare business proposal Produce relevant plans	Feasibility report Overall plan Technical and resource plans for next phase User approval to proceed
3 Systems Analysis	Analyse detailed user requirements and determine acceptance criteria Produce relevant plans	User specification Acceptance criteria Installation strategy Education and training plan Technical and resource plans for next phase User approval to proceed
4 Systems Design	Design system from user specification Produce relevant plans	System design Development plan Testing plan Technical and resource plans for next phase User approval to proceed
5 Programming and Testing	Design, write and test programs Design procedures and forms Complete documentation Complete systems documentation Produce relevant plans	Programs procedures Systems documentation Education and training documentation Technical and resource plans for next phase User approval to proceed
6 Acceptance Testing	Prepare and conduct acceptance tests Produce relevant plans	User accepted system Technical and resource plans for next phase User approval to proceed

Stage	Activities	End Products
7 Implementation	Prepare for conversion Convert system Monitor initial running of system	'Live' accepted system
8 Post Implemen- tation Review	Analyse results of system monitoring	Final report to project board

When structured methods are used, the development phases are changed in content and output and the amount of user involvement is significantly increased. The phases shown below are only an overview of the structured approach. It is covered in more detail in Chapter 15.

Phase	Activities	End Products
1 Feasibility Study	Prepare business proposal Produce relevant plans	Feasibility report Overall plan Technical and resource plan for next phase User approval to proceed
2 Analysis	Analysis of systems operations and current problems	Data flows and data structures of current system Problem and requirements list
	Specification of required system	Requirements specification, including: Required data flows, Required data structures, Required entity life histories
	Selection of user options Produce relevant plans	Specification of selected option Technical and resource plans for next phase User approval to proceed

Phase	Activities	End Products
3 Logical Design	Produce data and process design for required system Produce relevant plans	Logical system specification Technical and resource plans for next phase User approval to proceed
4 Physical Design	Map logical onto physical design Create development plans Formalise manual procedures Create program specifications Create ops instructions Create system test plans Create file/DB definitions	Physical system specification Technical and resource plans for next phase User approval to proceed
5 Implementation	Code and test programs Carry out link and system tests Prepare and conduct acceptance tests Produce relevant plans	User accepted system Technical and resource plans for next phase User approval to proceed
6 Changeover	Produce user manuals Carry out file/DB conversion Monitor initial running of system	User manuals 'Live' accepted system
7 Post Implementation Review	Analyse results of system monitoring	Final report to project board

Irrespective of the development method used, the work of analysis design and implementation will need to be planned and controlled. The simplest project planning and control cycle is:

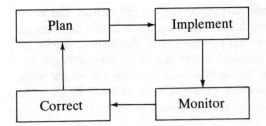

A more realistic representation for a computer based application development, however, might be:

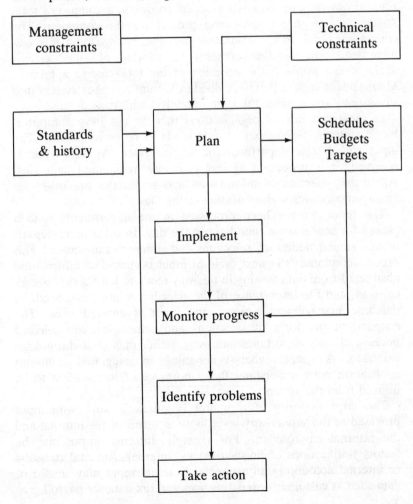

Management constraints determine the overall budget and time-scale within which the work must be done, and needs will change during development necessitating changes to the system. Technical constraints will include changes to the hardware and software environment within which the system will operate. A much more detailed treatment of project planning and control is given in *Systems Project Management*, also published by Pitman.

1.3 The design process

For various reasons, it is not practical to design a company's total system as one major short-term project for the computer. The company cannot accept a massive one-time change, and management needs time to define problems. The shortage of skilled systems analysts also prevents the adoption of the total-change approach. Management should therefore divide the total company system into subsystems and provide the systems analyst with their details.

In most commercial organizations that do not have computer-based systems, throughput of data tends to be chronological. The input to any one department occurs in cycles. As each cycle is received, it is processed against manually maintained data files. Additions, selection of information and summaries are made for management control and updating of the files.

The input, having been processed in the department, is then passed for action some time later in the day. In subsequent departments, similar additions, selections and summaries are made. This process continues; the next cycle of input occurs. The information abstracted from data flowing in this way tends to have a chronological bias, and the importance of the selection is not recognized. In database management systems, the input is entered once. The outputs, in the form of selections and summaries, are selected because of their importance and usage rather than on a chronological basis. A systems analyst designing an integrated computer application must understand the importance of information to be utilized from the system.

The large majority of computer installations work with input provided at the supervisory level, both in terms of the internal and the external environment. For example, internal output may be factory notifications of finished stocks, interdepartmental transfers or internal accounting adjustments. External input may consist of data such as customer orders, invoices and tax data for payroll.

If the computer is used at this level with these inputs, its outputs can provide the next level of management with information such as historical analyses, summaries and internal file data. The outputs provided to the next level of management may be similar to the inputs, such as invoices, statements and cheques.

If a company intends to use the computer to perform tasks that will directly help management to form policy, the inputs to the system must enable the computer to provide probability statistics. At this level, the internally provided company inputs will consist of as much of the company's history, including all aspects of the total system, as can be accumulated. External factors about economic trade and local conditions must be added to make a comprehensive management information system.

In practice, the areas in which a computer is employed are never as simple as those described above. It is quite conceivable that a microcomputer can provide top management with some information, and that a large computer will handle the day to day data processing work of the company. However, the help a computer can give management is only that of a tool whose main function is to provide feedback from the company's activities. Management can then establish much closer control over the future use of company facilities. Computer outputs provide the company policy-makers with much information, so that a more coherent plan of action may be passed to the chief executive.

After the management requirements of a system are established, the analyst must generate a set of descriptions to explain how outputs are derived from inputs. Management requirements will normally be output oriented. They will have to be refined, and further outputs will need to be determined.

The analyst may be involved in determining the inputs that are required. Once input information is captured by a system, it must be preserved in ordered files of records. The analyst must determine what files are needed and what records should be kept. In order to obtain the required output the mechanical and clerical procedures that will be required should be specified.

The design sequence is therefore as follows: (1) output (results), (2) inputs (data), (3) files (files), (4) procedures (program). Stages (2) and (3) may have to be resolved simultaneously. Similarly, stages (3) and (4) are interlinked.

In general, this simple sequence must be followed several times. At each stage, decisions that will affect subsequent stages are made. It may be necessary to change decisions made in earlier stages for reasons of technical feasibility, programming complexity, operating cost and so forth. Thus the process is one of successive repetition of the design procedure until a satisfactory solution to the overall problem is found.

Although a major part of the system is defined by the computer programs that are used, well annotated programs give only an understanding of the programming mechanics. The system specifications are absolutely essential; they alone constitute the definition of the system. As such, they are the major means of communicating the project's requirements to the programming and operational staff.

It was stated above that the design process is iterative; that is that successive refinement of the design is required to generate the final acceptable solution. This implies that at each stage of development the design must be examined to check its adequacy. It will ultimately be judged by the user (management) and by the chief systems designer, if present, who will also want to assess the adequacy of the design.

The structured analysis and design methods introduced in Chapter 15 greatly simplify this process of design validation and ensure that a high quality solution is produced at every stage.

The criteria of good system design may be summarized as follows: (1) management objectives realized; (2) well defined computer system; (3) well designed human aspects; (4) efficient and timely operation; (5) carefully planned and tested implementation; (6) accurately estimated costs; and (7) rigorous design methodology. The various aspects of this overall process are discussed in the following chapters. A careful application of the principles shown will help the analyst to produce good systems design consistently.

1.4 Human aspects of the job

The analyst must constantly remember that systems involve and affect people and must anticipate a variety of possible reactions to the system and the reasons behind them, even if they seem irrational.

We all have our systems of belief that cause our viewpoints to differ. They are based upon our experience, education, emotional make-up, intelligence, knowledge of specific areas, and our particular interests. Highly intelligent, logically minded people may have radically different yet reasonable views on the same subject. Using the same data, they have reached these views because their basic premises differ. These premises have been developed through the years, and are based on assumptions about the many non-measurable factors in our environment.

Few people are both highly intelligent and coldly logical. Our emotions are close to the surface and have a profound effect on our thinking. We tend to believe what we want to believe. Some of the illogicalities that are normal include generalizing from particular cases, assuming cause and effect from a correlation, attributing a logical reason to an emotional belief, and transferring subconscious dislikes to people or things our conscious minds will accept as scapegoats.

The main motivating forces in mankind have been identified as follows: (1) physiological needs such as those for food, shelter, rest, variety, safety and sex; (2) safety and security against danger and loss of physiological satisfactions; (3) social needs such as the needs of association, acceptance by associates, giving and receiving friendship; (4) ego needs such as recognition, sense of achievement, prestige, independence, realizing one's own potential and being creative. Other motivating factors are curiosity, which often causes people to look for change or new ideas, and habit persistence, or preferring the familiar to the new. Habit persistence is associated with problems of unlearning. It is difficult to change automatic reactions to external stimuli.

Most people have achieved a reasonable degree of satisfaction in meeting the above motivating needs. Fear of losing them provides powerful resistance to change. Helping people meet these needs has been a basic approach in motivation research and salesmanship for many years. However, most people resist this approach when they realize it is being used.

The analyst must persuade others so as to overcome resistance to change. As much information as possible should be gathered about the system of beliefs of the persons to be persuaded, and about the objections that may be raised. Careful observation, including attention to expressions and casual words, may help detect what someone is trying to conceal. If a group rather than an individual is concerned, the analyst should identify, and try to persuade, the most influential person in that group. The analyst should be established as a person to be trusted and able to persuade others out of sincerity and concern for their interests.

There should be a planned approach to determine intermediate objectives, considering all possible objections and the alternatives that can be offered. Timing should be such that the degree of pressure will get individuals to analyse their present situation or the products they are using and create a desire for change. The following suggestions may be useful during the investigation:

(1) Do not expect instant conversion.

(2) Induce participation in decision-making; this causes commitment to the decision.

(3) Avoid using too many arguments by emphasizing essentials.

(4) Ask questions that will emphasize areas of agreement.

(5) Aim at a mutually satisfactory solution, not a total conversion.

(6) Avoid any criticism of the past; concentrate on positive aspects of the change and a common desire to make progress.

(7) Listen sympathetically to problems and objections but do not assume that verbal objections are necessarily the real ones; there may be rationalizations of emotional objections that the person knows are irrational and will therefore not admit.

(8) Present alternatives, such as 'If you buy one would you prefer model x or model y?'

(9) Be wary of negative suggestions, such as 'It would be fine if we could do so and so, but it is not really possible in a firm like ours'. This approach may cause an aggressively favourable reaction to the implied challenge, but if it fails it creates a situation from which it is difficult to recover.

Assume that acceptance has been won and stress the advantages and benefits while admitting the difficulties. Suggest that management and supervision will be able to overcome the inevitable transitional problems and make the new system work.

Ask for suggestions as to how the more obvious problems may be overcome. Emphasize continued support and help both during and after the change. Once agreement seems reasonably certain, get the plans, including a program and time-table, accepted as quickly as possible. Give management and supervision the credit for the change. Remember that to be logically right can be psychologically wrong.

Some ground rules for resisting persuasion are worth considering, since they indicate what the persuader must overcome. These may be that the people resisting persuasion behave in one of the following manners:

(1) They don't listen.

(2) They attribute ulterior motives to the persuader, preferably behind his or her back to others who may be affected.

(3) They concentrate on disliking the analyst.

(4) They exaggerate objections, especially the danger of reper-cussions and the unsuitability of timing.

(5) They raise the temperature of the discussion, and discuss personalities wherever possible.

(6) They keep their real objections to themselves.

(7) They stick to their prejudices.

The analyst must consider the possible reasons for which em-ployees resist change. These are as follows: (1) fear of losing one's job, of wage reduction, of inability to learn a new job, of loss of prestige, of loss of interest in one's job; (2) suspicion of manage-ment's motives in making the change; (3) resentment against per-sonal attack, or a feeling that any change is a personal criticism of the way a job was being done; (4) social upset caused by breaking up a working group; (5) ignorance, or fear of the unknown.

Among the ways of overcoming these background reasons for resistance are the following:

(1) Keep people in the picture well in advance, give the full reasons, and sell the benefits.

(2) Give people an opportunity to participate by making sugges-tions.

(3) Give security, which may mean guaranteeing the financial future or providing retraining.

(4) Take time in introducing change; create a favourable atmos-phere; give people time to get accustomed to an idea before implementing it. (There are, of course, rare occasions where the quick introduction of change without notice may be best.)

(5) Provide sound personal examples. Much depends on the degree of confidence employees have in management.

(6) Cultivate the habit of change. If changes are frequent, people will be more used to the idea, and changes will be more readily acceptable.

It is also important to recognize that over a period of time the climate within which systems are implemented will change. During the 1960s systems were built to designs dominated by computing techniques and the need to manipulate all non-computer processes to suit machine processing. In the next decade systems designers tried much harder to meet user requirements and to provide the kinds of system that users needed and would welcome. Success was by no means uniform of course, and implementations were still met with hostility and rejection. The task for the 1980s is to eradicate this dissatisfaction. There are two major influences upon the climate within which this has to be done: first, technological change; second, social power changes.

The advent of the microcomputer increases enormously the scope and flexibility of computer systems. In addition, however, it seems to bring to life the very worst fears for employment prospects in the future. Second, there has been a shift in power from the top to the bottom of most organizations—not merely through a growth in trade union strength but also due to political and social forces outside employment.

Thus the implementation of computer systems has become much more of a collective, almost negotiating procedure with all affected parties represented. The use of 'new technology', whether micro or on-line screens, which seriously alters working methods is now no longer an implementation training task but a major event to be negotiated between management and workforce. Quite what happens when *management* tasks change and use is made of the new technology and who will do the negotiating then is not clear!

A new methodology based on participative systems design has been proposed to cope with these climatic changes. It is based on the belief that all employees have a right to influence the organization of their own work system. It is claimed that this increases job satisfaction since those whose job is about to change are better able to diagnose their own job satisfactions than is a systems analyst. Efficiency is also improved since employees in these situations of change are likely to know of day to day work problems and to have invented solutions to them. This methodology—although bringing new difficulties with it—clearly offers one solution to the problem of implementing change.

1.5 Personal qualities and training

The disadvantage of setting down a list of qualities required for any job is that no one person possesses them all. However, it is worthwhile to recognize the attributes that the job demands, since the trainee will probably acquire those that are lacking as his or her experience grows.

An analyst must be able to discover the fundamental logic of a system, produce sound plans and appreciate the effects of new facts in planning. The analyst must be perceptive, but must not jump to quick conclusions, be persistent to overcome difficulties and obstacles, and maintain a planned course of action in spite of setbacks.

There is also a need for stamina, strength of character and a sense of purpose; a broad, flexible outlook, an orderly mind, a disciplined approach and logical neatness are essential. The job will frequently require working without direct supervision and the ability to express thoughts, ideas and proposals clearly, both orally and in writing. To maintain control through numerous interviews requires an accurate and precise conversationalist who must have more than average social skill to communicate and work with others.

Suitable candidates for systems analysis trainee positions often come from a background of analytical work in the commercial area or computer-oriented work involving programming. Since the systems analyst will need both types of experience, the first step in training must be to develop the area of experience that the trainee lacks. Some trainees must develop communication skills, both oral and written; others must be trained in application areas. The level of experience to be developed may vary from a restricted and detailed knowledge of a specific area such as steel production to an overall understanding of a broad area such as management information systems.

This area of training is similar to an apprenticeship and cannot be satisfied, as has sometimes been tried in the past, by a two week course in business organization. The more successful analysts have served lengthy apprenticeships in an application area in addition to their professional qualifications.

The analyst must also be trained in the use of such relevant hardware as computers and peripheral equipment and must know the types of equipment best suited to the solution parameters, overall methods of computer application strategy, and detailed computer systems techniques. Training in this area must be continuous, because new equipment and techniques arise constantly.

Finally, the analyst must be trained in an appreciation of software and know which packages are currently available, so as to decide which are suitable for the system's needs. There is also a need to understand the principles of programming. It is pointless to specify a system that cannot be programmed with currently available software.

2 Systems Investigation and Analysis

This chapter deals with the work of the analyst in fact-finding, fact-recording and analysis so as to be able to describe in documentary form the essential operational requirements of a business system to satisfy the user's needs. Out of this work comes a definition of the outputs required from the system, the input data needed to generate the output information, the processing requirements and the operational objectives.

Systems investigation falls within the early stages of systems work and is part of what is commonly termed the system life-cycle. This is subdivided into the following stages: (1) analysis, (2) design, (3) implementation, (4) maintenance and review. Each one of these stages may be subdivided and have a number of tangible products. The following is an example of the way in which the stages of the system life-cycle may be used by a typical business user:

Stage	Subdivision	Product
Analysis	Preliminary study	Feasibility report
	Detailed study	User-system specification and report
Design	Detailed design	Detailed system specification and report
Implementation	Computer programming	Programs
	User-system procedure design	User manuals and training
	File conversion/creation	Specification and plans
	Testing and trials	Specifications, plans and results
	Changeover	Plan
Maintenance and review	Post-implementation review	Report
	Audit	Report

The stages of the system life-cycle form the basic problem-solving methodology of systems analysis. First, the problem is analysed, then a solution is designed; next the solution is put into practice and finally a review is undertaken to see how well the solution in reality solves the original problem as defined. This is a practical approach to solving business problems.

Systems investigation may be undertaken by the analyst as part of a feasibility or preliminary study. In this kind of work the analyst may find he is working alone, or in a very small team, in close collaboration with user management. Such studies may be strategic and conceptual in nature, covering large areas of an organization's data processing requirements, or they may be smaller studies of a limited operational area. In both cases the preliminary study is generally undertaken to get some idea of whether it would be worthwhile investing additional resources in more detailed studies.

On the other hand, systems investigation may be undertaken by the analyst as part of a detailed study, the outcome of which will determine what, when and how information will be produced by a new system. It is vitally important that this work is thoroughly done — and done to the satisfaction of management, the end-user and the computer services department.

2.1 The origin and objectives of systems investigation

Normally a systems investigation will be governed by terms of reference drawn up by a steering committee or by the line manager requesting the investigation. Oral terms of reference are unsatisfactory and if given such terms of reference the analyst would be well advised to prepare a draft based upon his understanding of them and present it to the originator for agreement.

Good terms of reference will define the scope of the investigation, e.g. which areas of an organization's activities are to be investigated; the objectives, e.g. to improve customer service (in a quantifiable way); the constraints, e.g. improvements in a system may be required to be operational by a certain date; the resources likely to be available for the investigation.

2.2 Planning the investigation

In the light of the terms of reference a plan will be produced to show how the investigation will be conducted — the methods to be used and the time-table of activities. One analyst may be given the entire responsibility for a small investigation whereas on a larger job several analysts will share the work.

After the planning and staffing are completed each department to be investigated must be consulted and agreement reached about the proposed general conduct of the investigation. Only when this has been obtained can the analyst begin the tasks of fact-finding, fact-recording and analysis.

2.3 Fact-finding

Fact finding begins at the analyst's desk with a complete and thorough review of all the information that already exists about the area to be investigated. This will include the Feasibility Study and the Terms of Reference of course, but should also include all existing materials that describe what happens now — or what is thought to happen now. Essential requirements will be organisation charts, together with job descriptions and stated departmental objectives. Don't be surprised to find that these are inadequate or non-existent. Descriptions of previous studies or of clerical procedures will be very useful if they exist. Tackling a fact finding assignment without having done appropriate preparation will lead to problems for two reasons. Firstly, users will realise that you are an ignoramus and haven't bothered to prepare for your assignment and deduce that you don't care about it, them, their problems or the new system they would like. Secondly, without adequate preparation you will lose the sharpness that comes from knowing the key questions to ask.

Various techniques are available to the analyst during fact finding. These include:

(1) Interviewing
(2) Questionnaires
(3) Observation
(4) Record sampling

These techniques are not mutually exclusive; indeed, in practice more than one technique will be employed to establish the facts. During the course of an interview, records may be inspected, a questionnaire completed and notes made about working conditions.

2.4 Fact-finding: interviewing

Only practice will perfect the art of interviewing; nevertheless, there is much that a junior analyst can learn from the experiences of others, and what follows are guidelines for the inexperienced — although some experienced analysts would benefit from an occasional reference to the fundamentals of good interviewing!

Interviewing is probably the most productive fact-finding activity for an analyst. Not only do interviews provide the facts, they enable the analyst to verify facts and they provide an opportunity to meet and overcome user resistance. Success will depend upon the interviewer's ability to work round, or avoid, resistance that may develop on the part of those staff whose work habits will be altered by any changes that may be recommended.

Obviously, therefore, a certain amount of skill is required on the part of the interviewer if these requirements are to be met. He should be impartial, tactful and, more positively, have skill in influencing others to accept advice they would prefer to ignore. Interviewing is the process of obtaining information — without upsetting the other party — by means of conversation; this entails being a good listener, being adept at keeping the ball of conversation rolling and being able to keep the subject on the right lines. Conversation is itself an art relying not a little upon the ability to suit the treatment of any subject to the person, the place, the mood and the moment.

The interviewer must bear in mind that he will be dealing with a very wide variety of personalities, and not only that but personalities at all different levels of authority. In a small department he may call on its head; in a large department the level at which he opens negotiations may be lower. But no matter whom he encounters, he will need to adapt his approach, timing and phrases to suit the particular person. It is also a common happening that the man in charge will wish to accompany the interviewer on his round of the staff. Sometimes, indeed, the head will attempt to answer all questions himself, or, on the other hand, try to conduct the interview! This situation demands much tact from the investigator, but it is essential for him politely and firmly to retain control of events.

Resistance to change will often produce a somewhat frigid climate for the interview and some extra care may be needed to produce a thaw. The more effective the job performed, in that satisfactory results are being achieved, the less likely are those engaged upon it to welcome someone 'fooling around with the methods they are using', which is one of the attitudes encountered. Yet it is often the case in this sort of work that someone with an impartial approach and a good knowledge of method can make improvements and secure economies in any operations, however successful they may already be. Those genuinely interested in the results they are achieving often become wedded to their well tried methods, are suspicious of change and may never have considered, or heard of, various alternative machines or processes. Systems investigations are concerned with existing procedures and well established habits, rather than a fresh system to be thought out independently from first principles. As a result, until a first-class reputation for helpfulness and soundness of advice has been earned, the systems man is met by some show of suspicion, by reluctance to change and sometimes by resistance to any suggestion of interference with 'my job'.

In systems work the interviewer works at the other man's desk, often in an open office without any privacy, and in full view and within the hearing of a number of other people. These other people may well be listening to what is happening and may themselves be interested in and affected by the conclusions reached at the interview. This lack of privacy heightens the risk, and makes the interviewer's own approach of greater importance than, for instance, that in a personal interview. But slips can be retrieved, and admission of error is in some cases preferable to incomplete concealment. Errors may be of facts and their misinterpretation due to ignorance or accident; or they may result from some personal misunderstanding.

The success of the interview itself is conditioned by at least two variable and probably incalculable factors—the personal prejudices of the two people affected, the interviewer and the interviewed. Of course, the interview does not stand alone as a means of fact-finding, but is used in conjunction with observation and examination. On the other hand, the interview is usually the only means of finding out something about that part of the job which cannot be seen, that which goes on inside the operator's head. And the higher the interviewer goes into an organization the more important is this side of the work. Therefore, to achieve the best possible results, it is worth his reminding himself beforehand of the need to eliminate his own bias. He must not have a pet solution, or one that is pat, but must be impartial and thorough. Second, he has to think about the person whom he is interviewing and put himself in the other's place. Many people are very enthusiastic and wholehearted about their job and may regard a visit by a systems analyst as a suggestion that their own work is not well done, and take this impression to heart. The interviewer must recall his own feelings when having his work critically inspected and allow this to make his approach sympathetic.

The two factors—that the interviewer is on the other man's ground, and is up against considerable resistance, actual or potential —mean that any art or tact that may lie in conducting an interview is worth considering. This will supplement the more definite process of studying what is actually done, what questions should be asked to complete the story already discovered by watching the job, and what meaning to give to the observations made or the answers received. To achieve this it is necessary (1) to make the interview impersonal; (2) to keep it objective; (3) to prepare for it beforehand.

When studying work methods it is necessary to disclaim any association with grades and wages; to confine questions and interest to the actual operations being carried out; and to avoid comparing one person with another, or asking anything that in any way reflects upon the person being interviewed. If possible, the investigator should encourage the operator to go through the motions of the job in detail so that he can get a factual picture of what is involved before listening to any account, for this is likely to contain a certain amount of opinion based upon how the procedure appears to the person doing it.

The wider the interviewer's experience, the greater becomes the temptation to recognize a problem at a glance. In fact, very few problems are exactly similar, and the discovery of some apparent cause that he has suspected to be operative is easily made. Hence the value of a strict self-discipline or freedom from bias, so that he refuses to guess, keeps his experience chained up, and does not prejudge the issue; when the stage of collecting facts has been passed, experience is needed to work on what has been found.

The strategy of the interview is contained in the advice to approach it fully prepared. The investigator should know the names of the persons in charge and those with whom he is going to talk. He should know what part this particular job represents of the work of the company, department or section, its purpose, and if possible, he should have found out any reason for the selection of the particular person to be interviewed. (Is he or she outstanding or average, easy or difficult?) He should also have ready a number of points upon which information is needed. On occasion, he may be able to formulate the actual questions required to gain the information he needs, although the words to be used are best left to come naturally. This will allow for the choice of expression to accord with the circumstances of the interview and the attitude of the operator, and will make his approach flexible. He must at all times think carefully, choose his ideas and words and reject any that may offend. Later, during the stage of observation of the process, it may be possible to arrange questions in a logical order so that each follows naturally from the last, so that the proceedings do not resemble an inquisition. This appeals to reason and helps keep the interview impersonal.

A subsidiary part of the preparation is that of making arrangements for the visit: finding out, for example, whether a general announcement has been made about it, and if so, in what terms, or whether suspicion has been created by what has been said or left unsaid. The interviewer may be forewarned by finding this out. If no announcement has been made, he must ask himself whether he would like to announce himself or whether he would prefer the person in charge (usually the section leader) to perform the introduction. Some people consider an introduction from 'above' essential. Again some people consider that the initial interview should be on the operator's home ground. These, together with an enquiry to ascertain whether suitable desk or working space has been arranged, are considerations which will help smooth the path of any interviewer.

Following these preparations and the initial introductions, the first stage in the interview is to state the reason for the visit. If possible, the interviewer should also mention the method he is following. Finally he should remember the following points:

(1) *Words* Use simple terms. It is likely that the interviewer and the person being interviewed will use the same words but will have quite different meanings for them. Apart from this, it is essential to ensure that words which will not be understood generally are not being used, and that the interviewer's words have conveyed the meaning he wishes them to convey. If necessary, he must repeat the same thought in different words and notice whether it registers.

(2) *Atmosphere* While keeping the discussion strictly impersonal, the interviewer must try to enlist the interest and co-operation of the operator in the assignment. If possible he should say that he is trying to save any unnecessary effort, to make the organization work more smoothly, to save time and improve the service. Most people are interested in at least one of these objects, and some in all of them.

(3) *Compliments*   Sometimes during an interview it is possible to give appreciation, and this should be done, but it is most dangerous to approve what may later require to be changed. Appreciation should be given rather for method than staff. However, the interviewer should never criticize the staff and avoid anything which will diminish the sense of importance of the operator, which may already have been threatened by the enquiry. This quality may be restored by the investigator confessing that he has no first-hand knowledge of the job itself and is therefore at a disadvantage until he has been shown what is done.

Throughout the interview regard should be paid to the small details of manner and politeness: for instance, it is advisable to ask permission to smoke; to arrange for both parties to be seated and comfortable; to cause a minimum of disturbance to the operator's work without forcing the pace or skimping the job. Neither will a show of politeness compensate for supercilious questioning or arbitrary interruptions. Affectations are unpleasant, and nervousness is catching. A straightforward and simple attitude toward the operator is best at all times. The balance of formality and informality differs with everybody, but it may be as well to start fairly formally, especially if either of the parties is nervous. When nervousness has departed the interviewer may get more information through informality, but he must remember that the operator may have had very little experience of being interviewed and may need help in explaining his work. The interviewer will have not merely to study his own approach, but also to guide the person interviewed without asking leading questions that suggest the answer to be given.

When the interviewer has gathered all the information he requires, it is not sufficient merely to say 'thank you' and leave; he may have to return at a later date, and in any case, must make sure that his enquiry has not upset the staff or caused any ill-feeling or difficulties for supervisors. It is necessary, therefore, to leave enough time to establish a friendly and helpful relationship. The interview should not run too long and waste time; on the other hand, it would be fatal to cut it short and thus hurt feelings. There is a need for keeping control of the interview; for a return to formality as opposed to a show of firmness. Experience alone will teach how to weave a little authority into a free co-operative discussion so that the interview may be ended on the right note. It is important to recapitulate very briefly the object of the interview and read any notes that have been made, in order to clear the suspicion of snooping. This should refresh the other man's memory of his earlier statements and give him a chance to correct himself. The interviewer may be well advised to ask then whether anything has been missed, which may encourage him to bring forth a valuable suggestion not previously mentioned.

It is also at this final stage that the operator may want to divulge 'unofficial' opinions or facts. He or she may have been looking forward to a chance of saying what the organization looks like from underneath, what time is wasted, and how stupid some procedures appear to those who have to carry them out. Even though this commentary rambles, it is as well for the interviewer to listen carefully for ideas which, however remotely, may bear on his assignment, whether they concern the organization, the management, the supervision or the work. He must not, however, unduly encourage such confidences or get involved in personalities by agreeing or commenting, but must merely listen.

It is obvious that no hard and fast rules can be laid down for interviewing. The possession of tact is a great asset. Yet there may be occasions when the interviewer has to stand his ground and point out that his duty is to get certain information and the other man's duty to the company is to supply it. Should this fail a 'showdown' would be a last resort, but only when tact has failed. Interviewing is perhaps the expression and use of this quality, whereas a showdown is an expression of force, reactions to which may be unexpected and unpleasant.

During an interview it is necessary to be able to distinguish between facts and opinions. Both are necessary, but their treatment will differ at the analysis stage. It is also important to ensure that both sides of a story have been heard, and for this reason it is always good practice to interview both the recipient and the originator of a document when possible.

The final important task in an interview is to ensure that all questions are asked at the correct level. Asking the wrong level of question or having inadequate interviewing ability does harm by wasting people's time. Examples of the subject matters appropriate for different levels of management and staff are as follows:

(1) *Top management* Policies, strategies and general pointers to the future directions in which the organization may move; assessment of commitment to change; extent of likely management involvement in project control; overall constraints associated with money, manpower and time-scales.

(2) *Line management* Departmental objectives, organization, procedures; information requirements; bottlenecks, problems;

labour constraints. Most managers will have suggestions for improving the effectiveness and efficiency of the operations for which they are responsible. Their guidance must be sought on the practicalities of fact-finding and the methods to be employed; whom and when to interview.

(3) *Operational staff* Functions actually performed; forms used; lines of communication; volumes of work; assessment of job satisfaction, morale; suggestions for improved procedures and forms.

The following check-lists will be helpful in preparing for and conducting interviews:

Planning the interview
    Clarify the problem and what is to be investigated.
    Find out as much as possible about the subject area.
    Prepare a list of main questions.
    Ensure authority to interview has been obtained.

    Choose the interviewees with care — especially if it is necessary to draw a sample.

Conducting the interview
    Ensure that the interviewee is at ease before embarking on the main fact-finding.
    Let the interviewee tell his story in his own words.
    Do not ask leading questions.
    Ask one question at a time.
    Keep the sequence of questions logical.
    Avoid the role of an 'expert'.
    Be straightforward and frank — avoid 'cleverness'.
    Ensure that answers are fully understood.
    Try to quantify vague references to percentages, etc.
    End the interview on a pleasant, constructive note.

After the interview
    Record the data at the earliest opportunity.
    Check the facts as soon as possible.
    Get agreement to the facts.

2.5 Fact-finding: questionnaires

This method of fact-finding must be used with caution; it is not as simple and effective as it may appear to the inexperienced analyst.

Care must be exercised over the form design; the form must be field tested and validated before its widespread use.

The kind of situations in which questionnaires may be the most practicable method of fact-finding are (1) where staff are located over a widely spread geographical area; (2) when a large number of staff are required to furnish data; (3) for verification of data found by other methods; (4) when 100 per cent coverage is not essential.

The basic design of a questionnaire falls into three sections: (1) *heading section*, which will describe the purpose of the questionnaire and contain the main references (name, payroll number, date, etc.); (2) *classification section*, which will contain data to be used for analysing and summarizing the total data (age, sex, job title, etc.); (3) *data section*, which will contain the data being sought.

One of the major problems in questionnaire design is that it is difficult to frame questions which are certain to obtain the exact data required. Also it is a feature of the bulk of questionnaire surveys that not all the forms will be returned—many people object to filling in forms while others delay completing them until they are eventually forgotten!

One advantage to be gained when a questionnaire is planned to precede an interview is that the respondent is given time to assemble the required information, thus saving his time and that of the systems analyst at the actual meeting.

The aim in questionnaire design should be to formulate the questions so that no misinterpretation is possible and no bias is possible in the replies, even though this aim is virtually impossible to attain fully. A covering letter may be necessary to explain the purposes of the questionnaire in order to avoid misunderstanding and to help gain co-operation. The date by which the questionnaire should be returned should be stated.

Sometimes a survey is needed on the distribution and usage of certain widely distributed report to ensure the inclusion of only useful information, to eliminate unnecessary recipients, to save their time, and to reduce the costs of preparation and distribution of the reports. A questionnaire reading as follows might be sent to each recipient of a report:

'We understand that you are on the distribution list for ..... Report. We are surveying the distribution of various reports to make sure they contain all the information needed; to eliminate unnecessary information; and to make sure that everyone who needs the information receives a copy of the report. We would like your help in this survey by answering the following questions:

(1) Do you still require this report?
(2) If you do, what information do you use?
(3) How do you use this information?
(4) What additional information do you think should be in the report?
(5) How would you use this additional information?

Please return your answers by . . . .'

If it becomes necessary to obtain a detailed story on the activities of all individuals in a section, a duties questionnaire may be used to collect the information. In the questionnaire each employee is asked to list his duties and the average amount of time he estimates he spends on each one. An example is shown in Figure 2.5.1.

XYZ CO. LTD — DUTIES LIST			
SURNAME AND INITIALS		DATE FORM COMPLETED	
YOUR JOB TITLE		DEPARTMENT	SECTION

Enter each main duty you perform, and indicate how many hours per week it requires—

No.	DESCRIPTION OF DUTY	Approx. hours per week
	Other activities (lunch, tea-breaks, etc.)	
	TOTAL HOURS WORKED PER WEEK	
	WHEN COMPLETED HAND THIS FORM TO YOUR SUPERVISOR	

Fig. 2.5.1   Duties questionnaire

In many instances, the estimate of average weekly hours cannot be very accurate. This should not present difficulties as the analyst knows that the use he will make of a questionnaire of this sort will not involve high accuracy.

It may be helpful to ask for the following additional information when considering duties: (1) a list of all forms and reports connected with the work, together with the information obtained and the operations performed upon them; (2) a list of all machines used; (3) a list of unusual, out of the ordinary, duties that seldom occur and therefore cannot be conveniently listed among the regular duties.

These 'duty lists' should be compared with any job description specifications which may exist.

2.6 Fact-finding: observation

Being able to observe an operation and to draw useful conclusions from the task tends to be an inherent ability that is very difficult to develop in those without it. To the unobservant, those possessing the facility seem almost to be gifted with a sixth sense. In practice, much depends upon powers of concentration.

An analyst will tend to make subconscious observations of his local environment during his visits to operational areas, and very small and unexpected observations may be significant at the analysis stage. For example, an analyst might be asked to investigate delays in production caused by the late arrival of subassemblies from the stores. During his investigation into the mechanics of the ordering system, he discovers that a spirit-duplicated release note has to be prepared to withdraw the goods from the stores and that the bottleneck is in the duplicating section. On further probing he is told by the supervisor of the section that this is caused by pressure of other work. However, during this interview he observes that the two duplicator operators seem to be spending most time on producing stencil-duplicated work. Subconsciously he notices that both members of staff are smartly dressed. After analysing the situation and deciding on the facts discovered that there is sufficient time and equipment in the duplicating section to enable them to prepare the necessary documents to schedule, he asks himself why there should be delays. A second visit to the section, based upon a theory evolved by his observations, elicits the real reason. Both operators consider the spirit duplicators to be dirty and, not wishing to spoil their clothes, they put off any jobs on the machines for as long as possible. The issue of appropriate overalls solves the problem and clears the production line delays.

The analyst may also be involved in undertaking planned or conscious observations when it is decided to use this technique for part of the study. It involves watching an operation for a period to see for oneself exactly what happens. Long periods should not be devoted to this method, however, as they are unnerving to those being observed and soporific to the observer. The technique is particularly good for tracing bottlenecks and checking facts that have already been noted. A related method is called systematic activity sampling, where observations are made of the state of an operation at predetermined times. The times are chosen initially by some random device, so that the operatives do not know in advance when they will next be under observation.

This latter technique is not to be confused with normal statistical sampling, although some of the ideas employed are common to both. The analyst may require to use statistical sampling methods when he is involved in a specialized form of observation—record inspection. As the name implies, this involves inspecting the actual results and records of the system under investigation. It requires detailed counting of entries, numbers of documents, timing of throughput, and so on, usually to qualify the facts obtained at interviews. Trends are important. For example, is the workload increasing or decreasing and if so by how much per month or year? Ratios can be significant as well. For example, what is the proportion of credit notes produced in a normal day's working compared with sales invoices?

The time factor will often prevent the analyst from making as thorough an investigation in this area as he would wish. In consequence, he must attempt to draw conclusions from a sample of the past and present results of the operation under study. This can be perfectly satisfactory if he understands the concepts behind statistical sampling, but is a very hazardous exercise for the untrained. One particularly common fallacy is to draw conclusions from a nonrepresentative sample. Extra care should be taken in estimating volumes and data field sizes from the evidence of a sample. For instance, a small sample of cash receipts inspected during a midmonth slack period might indicate an average of 40 per day, with a maximum value of £1500 for any one item. Such a sample used indiscriminately for system design might be disastrous if the real life situation was that receipts spanned from 20 to 2000 per day depending upon the time in the month, and that, exceptionally, cheques for over £100 000 were received.

Searching methods also aim to quantify the analyst's data about the existing system. Information can be collected about the volume of file data and transactions, the frequency with which files are updated and the accuracy of the data. An assessment of the volatility of the information can also be made. The usefulness of existing information can also be questioned if it appears that some file data is merely updated, often inaccurate or little used. All of the information collected by record searching can be used to cross-check information given by users of the system. Whilst this doesn't imply that user opinion will be inaccurate, discrepancies can be evaluated and the reasons for them discussed.

Observation check list
Observe

(1) *Working conditions*: light, heat, noise, interruptions

(2) *Layout*: ease of access, proximity to colleagues, filing systems, telephones, etc.

(3) *Ergonomics*: workstation arrangements for microcomputing, use of terminals, furniture layout, adequacy of furnishings

(4) *Supervision*: availability when needed

(5) *Workload*: light, heavy, variable, bottlenecks

(6) *Pace and method of working*: are there peaks and troughs of activity, does everyone follow the same method

2.7 Fact-recording

The only tangible product of the analyst's work during the investigation stage of systems work is the documentation produced by him. A professional approach to systems work requires a systems team to use standard documentation. High quality standard documentation offers the following advantages:

(1) *Aid to completeness* Incomplete forms speak for themselves; good standard forms ensure that all the right data are collected, recorded and cross-referenced.

(2) *Aid to analysis* Appropriate types of charts and tables enable the essential features of a system to be highlighted.

(3) *Aid to communication* Standard forms help all team members to communicate with each other in an unambiguous way. They facilitate the modification of systems designed in the past and ensure that systems currently being designed may be easily modified in the future. The facilitate communication between analysts, programmers, operators and users.

(4) *Aid to training* New entrants can become effective in a shorter time through common installation standard documentation. Transfers of staff from team to team are facilitated.

(5) *Aid to management* Until the system is actually up and running the only tangible product of the analyst's work is the documentation. Standard documentation enables the project manager and the analyst to agree on what is to be produced by what dates. A project can be subdivided into many minor stages by reference to the documentation to be produced.

(6) *Aid to security* The documentation of a system may be likened to an architect's drawings of a building; if the building is damaged or destroyed the drawings enable repair or reconstruction to be carried out as quickly as possible. Likewise the documentation of a computer system is indispensable for efficient remedial action in the event of a system fault or failure.

Thus, standard documentation is much more than a means of communication between analyst and programmer—important though that is. The discipline of conforming to standard documents is a sign of prefessionalism and also ensures that vagueness and incomplete ideas are dispelled.

The NCC Data Processing Documentation Standards provide a tried and tested set of standard documents which serve all the analyst's needs. Their scope covers the work of analysing, designing, implementing and maintaining business systems. A business system consists of people and machines—and the data and procedures used by them. A good documentation system records unambiguously all their relationships and the interactions between them.

No purpose will be served by good documentation if it cannot be retrieved easily and kept up-to-date—so there is the additional need for a good document referencing system. The documentation of a system may be referenced as shown on page 34.

In practice, during the normal course of an analyst's work charting techniques of one kind or another are found most useful for recording organizational structures, activities, procedures and flows of data, whereas standard forms are found most useful for recording information about data—their content, organization, relationships and volumes, etc. Some examples of these are now given.

Organization charts (*see* Figure 2.7.1) indicate the formal groupings of people in a department. They show the line functions and give some idea of the responsibilities of the personnel involved. Their value to the analyst lies in the levels of authority which they display. From a well drawn organization chart, it is possible to obtain an overall impression of the scope of the department under investigation before a visit is made.

An activity chart describes the breakdown of the structure of a department by specifying what each person does. A typical example is shown in Figure 2.7.2. It can provide the analyst with an indication of the man-hours involved in the throughput of work.

A procedure flowchart does what its name suggests: it charts the flow of information through a specific procedure, or process. It is not concerned with the responsibilities of the individual personnel performing the process, but only with the procedure itself. The analyst may never have seen this type of process before. The jargon of the system may be a foreign language, a language full of 'Pink Slips', 'Bar Numbers', 'Stop Cards' or 'Joe's Book'. He is in unfamiliar territory. A motorist in an unfamiliar country does not rely on a series of written or verbal directions; he buys a road map.

The systems analyst cannot buy a map. He has to make his own,

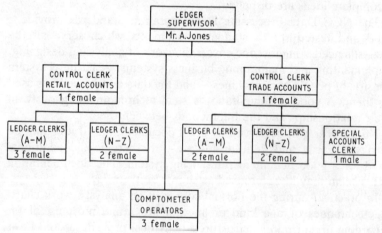

X.Y.Z. Co. LTD.—ACCOUNTS DEPT. ORGANIZATION CHART—SALES LEDGERS

Fig. 2.7.1 Specimen organization chart

Activity	Total Man-hours	Supervisor	Man-hours		Man-hours	Analyst	Man-hours	Clerk	Man-hours
Inventory control record and report	33					Sample check postings	8	Post orders	10
						Prepare inventory report	12	File posted papers	3
Requisition materials	23	Review and sign	6	Determine and separate cards requiring orders	6				
				Make pencil copy of requisitions	8				
				Maintain requisition control register	3				
Expedite shortages	20					Prepare and dictate expediting correspondence	10	Prepare and dictate expediting correspondence	10

Fig. 2.7.2   Specimen activity chart

and in making the map he learns all the features of his territory. An incomplete road map will show roads ending for no apparent reason. An incomplete procedure chart will show similar gaps in clerical routines. The message to the cartographer is the same in both cases—'Go and have another look'.

Procedure charting has three main functions:

(1) It enables the systems analyst to be reasonably sure that he has covered all aspects of the system.

(2) It provides the basis for writing a clear and logical report.

(3) It is a means of establishing communication with the people who will eventually operate the new system. It is at this stage that the user can be involved. He will be interested in seeing his own narrative appear on a chart in symbolic form and probably even more interested to find that he can understand the technique.

The second of these three functions forms the basis for the 'analysis' of the system. A system cannot be analysed until it is expressed in the form of a clear, logical report. Once the report is circulated then

Filing reference		Name/definition
1		Background (terms of reference, objectives, constraints)
2		Communications (information gathered concerning scope and purpose)
	2.1	Discussions, meetings
	2.2	Correspondence
	2.3	User manuals
3		Processes (information about organization, methods and activities)
	3.1	Overview of total system
	3.2	User—clerical subsystem
	3.3	Computer sybsystem
4		Data
	4.1	Clerical data
	4.2	Source data files
	4.3	Output data files
	4.4	Stored data files/database
	4.5	Source data records
	4.6	Output data records
	4.7	Stored data records/item groups
5		Facilities
	5.1	Accommodation
	5.2	Hardware, software
	5.3	Ancillary equipment
6		Tests (information about activities to prove the system design)
	6.1	Specification of test data and requirements
	6.2	Test plans
	6.3	Test operations
	6.4	Test logs
7		Costs
	7.1	Original cost estimates and system justification
	7.2	Periodic cost statements and variance analyses
8		Performance (timings, volumes and growth, etc.)
9		Documentation control
	9.1	Copy control
	9.2	Amendment list
	9.3	Outstanding amendments

analysis can begin. At this stage, everyone interested should be invited to comment and criticize; these people will do much of the systems analyst's work for him—if he cares to sit and listen.

Three main procedure flowcharting techniques are in use at present to represent activities and events symbolically. Each method employs its own set of symbols to signify main steps in the pro-cedure. The three 'standards' are (1) work study (ASME, American Society of Mechanical Engineers) symbols—listed in Figure 2.7.3; (2) horizontal form flow chart (HFFC) symbols—listed in Figure 2.7.4; (3) computer-based (ECMA, European Computer Manufacturers Association) symbols—listed in Figure 2.7.5. Each method has its advocates and use of one method as opposed to another is one of suitability or personal preference. The purpose of them all is to reduce a procedure to its basic component parts and to emphasize their logical relationships, so that a connected pattern of activity can be traced through from beginning to end. Such docu-

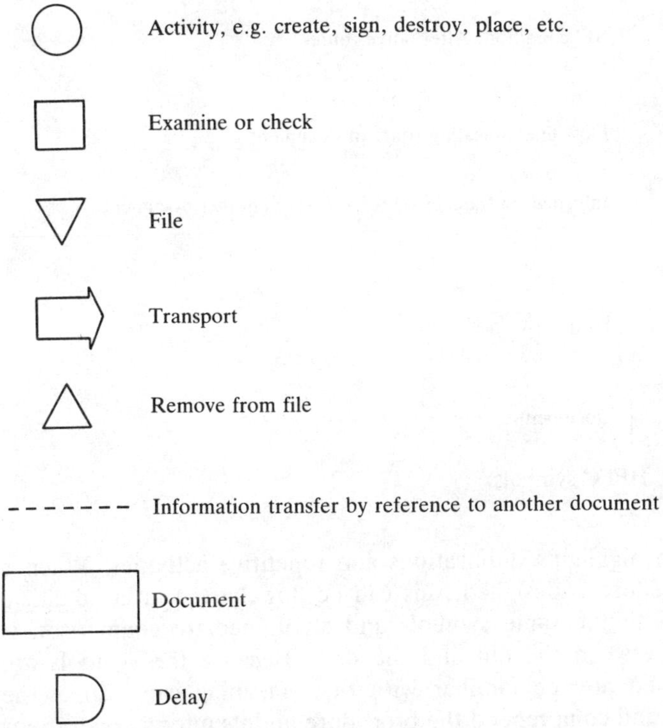

Activity, e.g. create, sign, destroy, place, etc.

Examine or check

File

Transport

Remove from file

Information transfer by reference to another document

Document

Delay

Fig. 2.7.3   ASME symbols

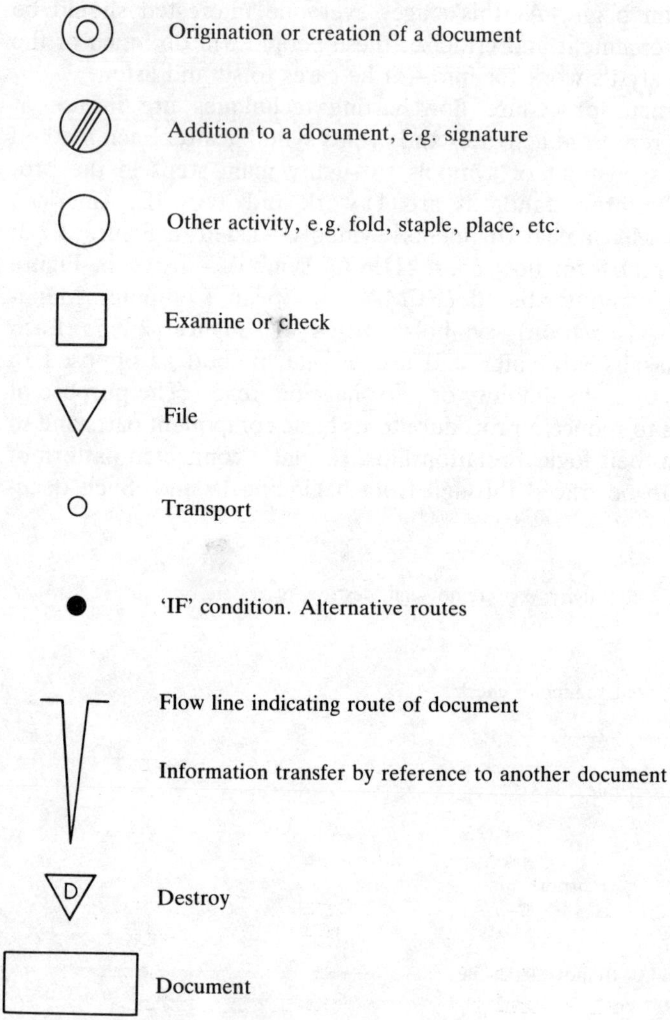

⊙	Origination or creation of a document
⊘	Addition to a document, e.g. signature
○	Other activity, e.g. fold, staple, place, etc.
□	Examine or check
▽	File
○	Transport
●	'IF' condition. Alternative routes
	Flow line indicating route of document
	Information transfer by reference to another document
▽D	Destroy
▭	Document

Fig. 2.7.4 HFFC symbols

mentation highlights duplications and repetitive activities. When a new procedure is designed, this can be flowcharted in an identical manner with the same symbols and an immediate comparison is possible between the old and the new. Because the symbols are standardized anyone familiar with their meaning can inspect the flowchart and comprehend the procedure and its internal operations very quickly.

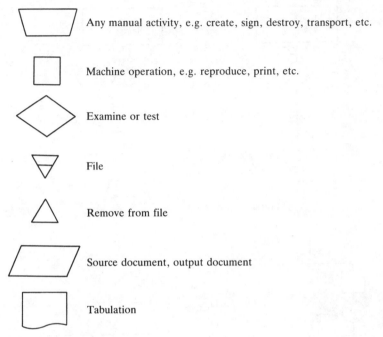

Any manual activity, e.g. create, sign, destroy, transport, etc.

Machine operation, e.g. reproduce, print, etc.

Examine or test

File

Remove from file

Source document, output document

Tabulation

Fig. 2.7.5 ECMA symbols

Figures 2.7.6, 2.7.7 and 2.7.8 give specimen flowcharts drawn to represent the same procedure using the three methods listed above to enable comparison between them. The procedure they partly illustrate, in narrative form, is as follows:

(1) The design draughtsman creates a design drawing. When finished he removes from file the Part Number Register and examines it for the last used part number. The draughtsman enters the next part number and part description in the register and the part number on the drawing and re-files the Part Number Register.

(2) The design draughtsman passes the design drawing to his section leader who checks for any errors. If he finds an error he marks it in red pencil and returns it to the draughtsman, who makes the necessary correction and re-submits the drawing to the section leader. The section leader re-checks the drawing and passes it to the chief designer for signature. If the drawing is correct when first checked the section leader passes it straight to the chief designer. The chief designer signs the drawing in the space provided and passes it to the Print Room.

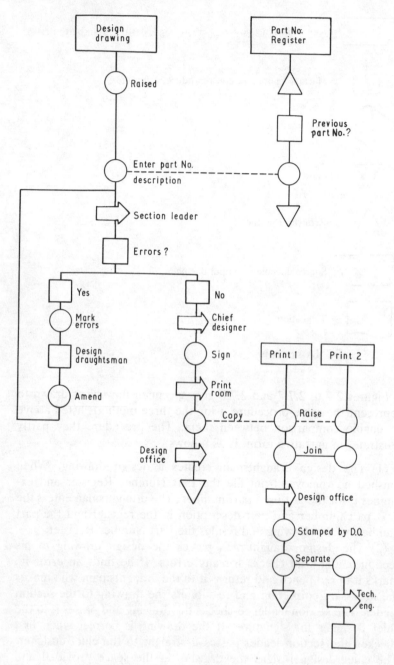

Fig. 2.7.6 Flowchart using ASME symbols

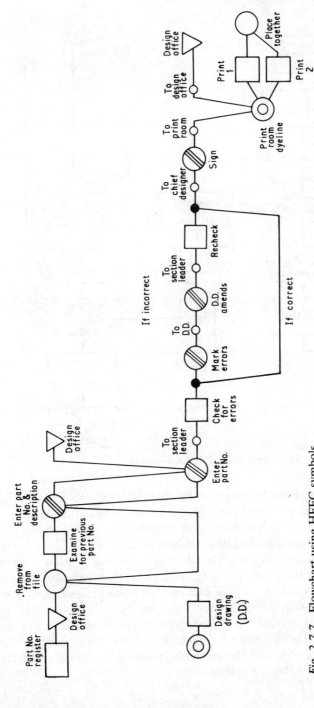

Fig. 2.7.7 Flowchart using HFFC symbols

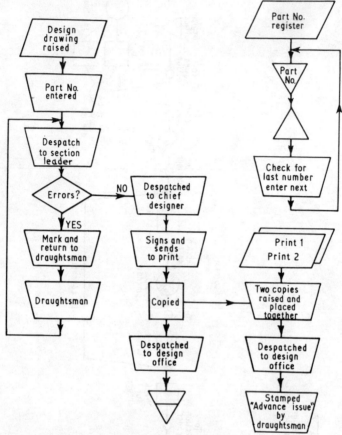

Fig. 2.7.8 Flowchart using ECMA symbols

(3) The Print Room prints two copies of the drawing on a dyeline printer and placing these together passes them to the Design Office. The original design drawing is returned separately to the Design Office where it is filed in the 'Master' file.

(4) On reaching the Design Office the No. 1 print is stamped 'Advance Issue' by the draughtsman who then separates the copies, files the No. 2 copy in the Design Office and passes the other to the technical engineer. The technical engineer checks to see if the drawing is the last for the assembly, if not he files it in a 'Pending' file. If it is the last, he withdraws from the 'Pending' file the remaining prints for the assembly, places them together and creates a Parts List (hand-written).

(5) The technical engineer, having raised the Parts List, sends it to the machines room supervisor who passes it to the punch operator. The punch operator punches a Breakdown Pack from the Parts List then machine sorts the pack into part number order within assembly level. The Parts List is returned to the technical engineer (Design Office) who files it.

(6) Having sorted the cards, the punch operator uses a tabulator to produce a complete breakdown Parts List. The cards are filed in the Machine Room and the Parts List passed to the technical engineer who then draws a complete set of No. 2 drawing prints from the file, places the whole lot together and passes them to the process planner.

The ASME and HFFC symbols are used mainly by work study and O & M practitioners; the ECMA symbols by computer specialists. None is ideal for charting both clerical and computer procedures. A set of symbols has been developed by the NCC to meet the needs of both clerical and computer procedure flowcharting. If used in conjunction with NCC standard documents they can be used to portray different levels of activity: (1) total system; (2) computer runs; (3) computer procedures. The NCC symbols and their significance are shown in Figure 2.7.9.

Good documentation standards will always (1) enable facts to be set out clearly in sufficient detail for the appropriate level; (2) provide means for comparison and analysis; (3) provide a method of referencing and filing to facilitate easy retrieval of information.

It is a general flowcharting convention that flowlines go from left to right and from top to bottom; exceptions are indicated by arrowheads, such as the one entering symbol 5 (Figure 2.7.10). These are not to be confused with the triangular arrowheads (symbols 13 and 15, Figure 2.7.10) which indicate movement of information.

Where a document containing a lower level of detail is available, relating to one of the symbols on another flowchart, cross-reference may be provided to that document in the top stripe of the symbol. The Clerical Procedure Flowchart (Figure 2.7.11) is a lower-level representation of symbol 2 in the System Flowchart (Figure 2.7.10). This particular flowchart uses the columns to differentiate between the different documents and files used in the procedure. The same procedure could be drawn in a different way to emphasize department activities, or as a single main flowline to emphasize processes. The choice as to how the columns are to be used is one to be made

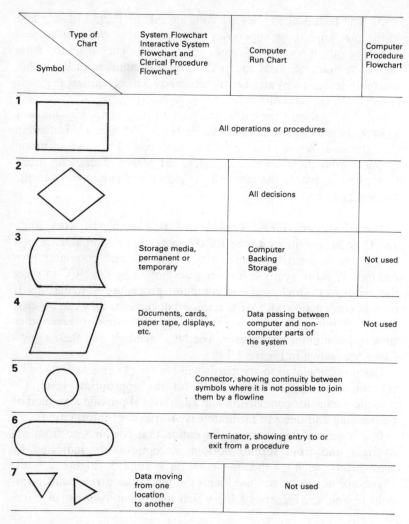

Symbol / Type of Chart	System Flowchart Interactive System Flowchart and Clerical Procedure Flowchart	Computer Run Chart	Computer Procedure Flowchart
1		All operations or procedures	
2		All decisions	
3	Storage media, permanent or temporary	Computer Backing Storage	Not used
4	Documents, cards, paper tape, displays, etc.	Data passing between computer and non-computer parts of the system	Not used
5		Connector, showing continuity between symbols where it is not possible to join them by a flowline	
6		Terminator, showing entry to or exit from a procedure	
7	Data moving from one location to another	Not used	

Fig. 2.7.9 NCC symbols

by the analyst, depending on which format brings out most clearly the essential factors. The broken lines (Figure 2.7.11) indicate relationships between different columns.

Other conventions of flowcharting are as follows:

(1) Crossed lines do not imply any logical relationship.

(2) Where two incoming lines join one outgoing line, they should not join at the same point.

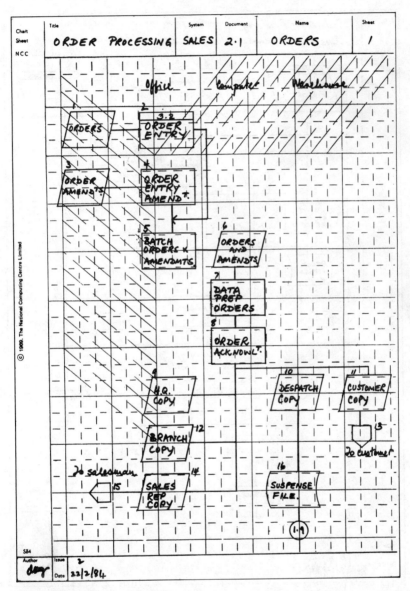

Fig. 2.7.10 System flowchart

(3) Annotation should be included within the symbol for 'process', 'decision', 'file' and 'input/output', giving an indication of the action, decision, file or document used. The annotation for movement (e.g. to pricing clerk, as symbol 8 on Figure 2.7.11) should appear to the

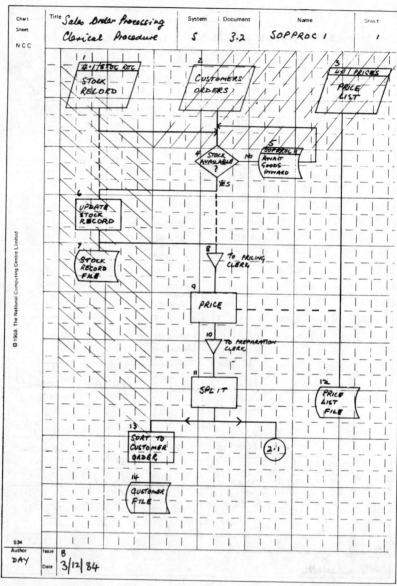

Fig. 2.7.11 Clerical procedure flowchart

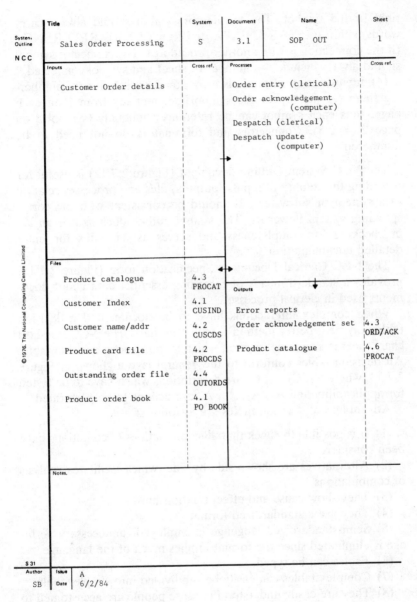

Title		System	Document	Name	Sheet
Systems Outline **N C C**	Sales Order Processing	S	3.1	SOP OUT	1

Inputs	Cross ref.	Processes		Cross ref.
Customer Order details		Order entry (clerical)		
		Order acknowledgement (computer)		
		Despatch (clerical)		
		Despatch update (computer)		

Files	Cross ref.	Outputs	Cross ref.
Product catalogue	4.3 PROCAT		
Customer Index	4.1 CUSIND	Error reports	
Customer name/addr	4.2 CUSCDS	Order acknowledgement set	4.3 ORD/ACK
Product card file	4.2 PROCDS	Product catalogue	4.6 PROCAT
Outstanding order file	4.4 OUTORDS		
Product order book	4.1 PO BOOK		

Notes.

S 31			
Author SB	Issue	A	
	Date	6/2/84	

Fig. 2.7.12 Systems outline

© 1976. The National Computing Centre Limited

right of the symbol. The terminator symbol should always carry words within it: START, EXIT, or PROCEDURE REFERENCE (if the connection is with another procedure). The connector symbol should always include the number of sheet and symbol connected.

(4) Identification of symbols on a flowchart is aided by numbering them (at the top of the symbol) sequentially from 1 on each page, thus each symbol can be referenced uniquely (symbol 5 on page 2 = 2.5). Connectors and terminators do not need to be numbered.

The NCC System Outline document (Figure 2.7.12) is useful for recording the totality of inputs, outputs, files and processes relating to a system or subsystem. It should be consistent with the corresponding system flowchart. The system outline document is an aid in checking for completeness and serves as an index for more detailed documentation.

The NCC Clerical Document Specification form (Figure 2.7.13) provides a standard means of recording essential data about documents used in clerical processes.

When complex interactions have to be documented a decision table may be a better method than a flowchart. The NCC standard chart sheet is a suitable form on which to construct a decision table. All decision tables conform to the general layout shown in Figure 2.7.14, where $C_1, C_2 \ldots, C_n$ are descriptions which have to be tested for applicability and $A_1, A_2, \ldots, A_n$ are actions to be performed.

Advantages of decision tables are as follows:

(1) It is possible to check that all combinations of conditions have been considered.

(2) Alternatives are shown side by side, which facilitates analysis of combinations.

(3) They show cause and effect relationships.

(4) They use a standardized format.

(5) Semi-standardized language is employed; unnecessary verbiage is eliminated since the format implies much of the language.

(6) Tables can be typed.

(7) Complex tables can easily be subdivided into simpler tables.

(8) They are easily understood because people are accustomed to using tables. Train time-tables, mileage tables on road maps, etc., are examples.

(9) Table users need not understand computers.

(10) Under certain conditions a computer program may be written directly from the decision table itself.

Clerical Document Specification	Document description Purchase Order	System POS	Document 3	Name PUORD	Sheet 1

NCC

Stationery ref. DS 46	Size A4		Number of parts 4	Method of preparation Typed	
Filing sequence By order number		Medium loose leaf binder		Prepared/maintained by HO Admin	
Frequency of preparation as required		Retention period 3 months after payment		Location HO Admin Supervisor	

	Minimum	Maximum	Av/Abs	Growth rate/fluctuations	
Monthly VOLUMES	20	300	120	no growth likely	

Users/recipients	Purpose	Frequency of use
- HO Admin	Raise order	daily
- Purchase accounts	To check against supplier invoice	monthly
- originator of order request	To check against delivery + authorise payment	weekly

Ref.	Item	Picture	Occurrence	Value range	Source of data
1	Supplier name		1 per order		POR
2	Item to be ordered	9(6)	5 per order	000001 - 999999	POR
3	Quantity of item	9(6)	as ref 2	000001 - 999999	POR
4	Est. cost of item	£99999.99	as ref 2	00001 - 99999	POR
5	Total order value	£9999 999.99	1 per order	000001 - 999999	POR
6	Delivery address		1 per order		POR
7	Authorised signature		1 per order		HO Admin
8	Date of order	99AAA 99	1 per order	valid date	HO Admin
9	Order number	9(5)	1 per order	00000 - 99999	pre printed

Notes

S 41

Author DM	Issue 3	Date 16.2.84

Fig. 2.7.13 Clerical document specification

Tables can be completed in one of two ways called *limited entry* and *extended entry* respectively. With a limited entry table, the statement of each condition or action is completely contained in (or limited to) the appropriate stub. The entry portions of this table

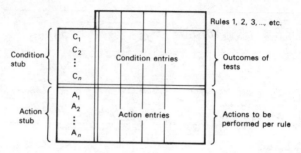

Fig. 2.7.14 Decision table layout

indicate only if a particular rule satisfies the condition or requires the action stated. Convention dictates that in the condition entry part of a limited entry table only three symbols are used: 'Y' (for yes) if the condition is satisfied, 'N' (for no) if it is not, and a 'hyphen' if the condition is not pertinent to the rule. Similarly, only two symbols are allowed in the action entry portion. These are 'X' if the action is to be executed, and a 'blank' if it is not.

In an extended entry table, the statements made in the stub portions are incomplete. Both the stub and the entry sections of any particular row in the table must be considered together to decide if a condition or action is relevant to a given rule. The advantage of the extended entry method is a saving in space, although an extended entry table can always be converted to a limited entry form. Figure 2.7.15 shows the two methods used to state the same problem.

Limited entry tables may be checked for completeness by a simple mathematical relationship. First count the number of blanks in each condition entry. Call this N, remembering that $N = 0$ when the there are no blanks. Then, for each rule in turn, calculate the value of 2 to the power of N. Sum the resulting values. This answer should equal 2 to the power of C, where C is the number of conditions given in the table. If the two answers do not agree, then either some rules are missing or too many rules have been inserted and the table must be re-checked for accuracy. If identical sets of conditions require different actions, then the table is said to be ambiguous, and again a check for this must be made. If different sets of conditions lead to the same actions, then the rules may be combined (eliminating redundancies).

There is one additional feature employed in decision tables, which makes the above checking method void. This is known as the 'else' rule. It is used when only some of the total number of possible

Limited entry	Rules				
	1	2	3	4	5
Is order valued between £0 and £10?	Y	N	N	N	N
Is order valued between £11 and £100?	—	Y	Y	N	N
Has customer satisfactory credit limit?	—	Y	N	Y	N
Approve order	X	X		X	
Allow discount of 3%		X			
Allow discount of 5%				X	
Refer to supervisor			X		X

Extended entry	Rules				
	1	2	3	4	5
Order value is more than	£100	£100	£10	£10	£0
Credit limit satisfactory	Y	N	Y	N	—
Approve order	X		X		X
Allow discount of	5%		3%		
Refer to supervisor		X		X	

Fig. 2.7.15 Limited and extended entry methods

conditions require testing. The required conditions are written as normal rules, and then a final rule having no conditions but entitled 'else' is inserted. The actions for this rule are normally to go to an error routine or to exit from the table. Insertion of this special type of rule is an instruction to perform only the tests on stated conditions. If all these are unsatisfied then the 'else' action is taken, without testing every possible rule (i.e. combination of conditions). Although this device can be useful in restricting an otherwise large table, it must be employed with extreme caution to ensure that no conditions which should have been tested have in fact been forgotten.

The problem to which the analyst must specify a solution will normally be written as a narrative description. The wording may be vague and conditions and actions are likely to be scrambled together. With experience, the analyst will be able to prepare a decision table directly from the narrative without difficulty. For beginners, however, the following method provides a systematic approach to the task:

(1) On the narrative, underline all conditions with a solid line and all actions with a dotted line.

(2) Enter the first condition on a blank decision table outline immediately above the double line, using extended entries.

(3) Enter the first action immediately below the line.

(4) Complete the table in extended form. Identify and consolidate similar statements.

(5) Check for ambiguity, redundancy and completeness.

(6) Insert 'else' rule.

(7) See if the table should be converted to limited entry by checking whether it will still go on to one page, whether any one existing entry will not extend to more than five new lines, and whether there are a reasonable number of entries on at least two lines.

(8) Check whether the problem would be better expressed on more than one table.

Figure 2.7.16 shows an original narrative, its underlined version and the decision table derived from the latter to illustrate the above points.

As a general rule of thumb, it has been suggested that a decision table should be considered when the number of rules multiplied by the number of conditions in the problem gives an answer of six or more. However, the tendency to draw up tables which are too large must be resisted; another general rule is that in a limited entry table of full size, the maximum number of conditions quoted should be four, which will give rise to 16 rules. It is stressed that these are rough guides only, and the analyst must use his common sense to detect when a table is too complex and would benefit from being split. Always remember that the aim is to communicate clearly and concisely.

An example of the limited entry decision table drawn to NCC standards is shown as Figure 2.7.17.

WHEN THE QUANTITY ORDERED FOR A PARTICULAR ITEM DOESN'T EXCEED THE ORDER LIMIT AND THE CREDIT APPROVAL IS 'OK', MOVE THE QUANTITY ORDERED AMOUNT TO THE QUANTITY SHIPPED FIELD THEN GO TO A TABLE TO PREPARE A SHIPMENT RELEASE. OF COURSE, THERE MUST BE A SUFFICIENT QUANTITY ON HAND TO FILL THE ORDER.

WHEN THE QUANTITY ORDERED EXCEEDS THE ORDER LIMIT, GO TO A TABLE NAMED ORDER REJECT. DO THE SAME IF THE CREDIT APPROVAL IS NOT 'OK'.

OCCASIONALLY, THE QUANTITY ORDERED DOESN'T EXCEED THE ORDER LIMIT, CREDIT APPROVAL IS 'OK', BUT THERE IS INSUFFICIENT QUANTITY ON HAND TO FILL THE ORDER. IN THIS CASE, GO TO A TABLE NAMED BACK ORDER.

Underlined version

WHEN THE QUANTITY ORDERED FOR A PARTICULAR ITEM DOESN'T EXCEED THE ORDER LIMIT AND THE CREDIT APPROVAL IS 'OK', MOVE THE QUANTITY ORDERED AMOUNT TO THE QUANTITY SHIPPED FIELD THEN GO TO A TABLE TO PREPARE A SHIPMENT RELEASE. OF COURSE, THERE MUST BE A SUFFICIENT QUANTITY ON HAND TO FILL THE ORDER.

WHEN THE QUANTITY ORDERED EXCEEDS THE ORDER LIMIT, GO TO A TABLE NAMED ORDER REJECT. DO THE SAME IF THE CREDIT APPROVAL IS NOT 'OK'.

OCCASIONALLY, THE QUANTITY ORDERED DOESN'T EXCEED THE ORDER LIMIT, CREDIT APPROVAL IS 'OK', BUT THERE IS INSUFFICIENT QUANTITY ON HAND TO FILL THE ORDER. IN THIS CASE, GO TO A TABLE NAMED BACK ORDER.

Limited entry table		Rule 1	Rule 2	Rule 3	Rule 4
01	Quantity ordered ⩽ order limit	Y	N	Y	Y
02	Credit approval = 'OK'	Y		N	Y
03	Quantity on hand ⩾ quantity ordered	Y			N
04	Move quantity ordered to quantity shipped	X			
05	Go to release	X			
06	Go to order reject		X	X	
07	Go to back order				X

Fig. 2.7.16 Preparing a decision table from a narrative

Chart Sheet NCC	Title: Allocation of Stock to orders	System: SOP	Document: 3·4	Name: ALTAB 1	Sheet: 1

C = 6
A ≥ 10
A = 7

RULES

	1	2	3	4	5	6	7
End of transaction file	Y	N	N	N	N	N	N
End of stock file	-	Y	N	N	N	N	N
Stock record key < trans. key	-	-	Y	-	-	-	N
Stock record key = trans. key	-	-	-	Y	Y	Y	N
Required amount > stock	-	-	-	N	Y	-	-
Stock level > 0	-	-	-	-	Y	N	-
Update stock record				X	X		
Allocate stock to order				X	X		
Re-order stock					X	X	
Write daily movements rec.				X	X		
Write back order record					X	X	
Perform error routine	X				X		
Read next stock record			X	X	X	X	
Read next trans. rec.				X	X	X	
Close files & end run	X	X					
Go to ALTAB 1			X	X	X	X	X

S34
Author: RLM
Issue: A
Date: 2/1/84

Example of limited entry decision table.

Fig. 2.7.17 Limited entry decision table to NCC standards

2.8 Analysis

Paradoxically this activity is not well understood by many systems analysts! Nor is it commonly recognized as an activity in its own right. Too often, investigation to provide the facts is followed immediately by design of a system. However, an intermediate stage is desirable to analyse the outcome of the investigation in the light of the objectives. Systems analysis is essentially about identifying and defining business problems which are worth solving within the resources likely to be available. Without systems analysis there is a real danger of designing and implementing systems which do not meet the user's needs—simply because the user's needs were not adequately identified and defined. Clearly the logical time to do this is after all the facts have been gathered and recorded—and before design is embarked upon. However, it is true to say that an experienced analyst will be doing some of the analysis as his investigations are proceeding. Nevertheless, there is a need for all the analytical effort to be brought together, and this is probably best done if it is focused on producing a specification of the user-system requirements—in user terms. The aim should be to get agreement on all the outputs, inputs and processing requirements—but substantially divorced from any detailed consideration of how these are to be satisfied by any new system. What is being sought is a description of the logical system—as opposed to a description of the physical system which emerges from the detailed design stage of a project.

The remainder of this chapter describes what tools and techniques are available to assist the analyst in the task of analysis and in the documentation of the results of this activity. Good clear documentation is essential for two reasons: (1) to enable users to agree the findings, conclusions and recommendations; (2) to provide an unambiguous foundation for the design work ahead.

It is essential that all relevant data be recorded. This is not as simple as it sounds. Here are some of the things one needs to know about data for future systems design: (1) its description; (2) its 'picture' (that is, its composition and size in terms of alpha-numeric characters, etc.); (3) its owner; (4) its relationship with other data items; (5) its source; (6) the processes in which it is used; (7) security aspects.

In recent years data analysis has come to be recognized as a major activity in its own right. To assist this analysis the data must be collected together in a methodical way and the NCC Data Pro-

cessing Documentation Standards cater for this. Figure 2.8.1 shows
the form for data definition. The idea behind this form is that each
item of data should be uniquely identified and defined and, in the

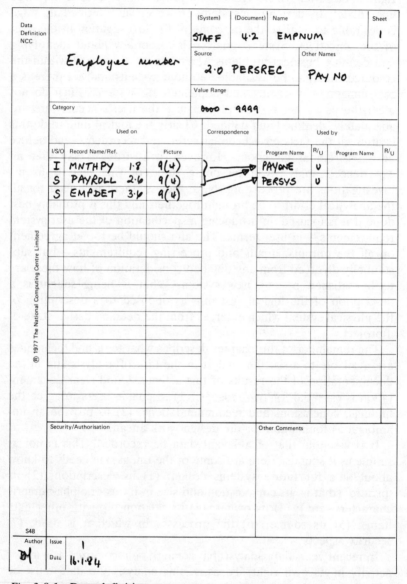

Fig. 2.8.1 Data definition

later stages of systems work, be cross-referenced to the records in which it appears and the procedures or computer programs which use it. The complete documentation forms a data 'dictionary'—an up-to-date reference manual defining all the data in a system. Such dictionaries may be created and maintained manually or by computer.

An example of the way in which the defined data may be analysed is the method known as 'normalization'. This has evolved out of the work of E. F. Codd into relational data-bases. Unnormalized data is illustrated in the first part of the example in Figure 2.8.2. Here the data describing an employee in a personnel records system is shown, the training courses' data is repeated for each course attended. The record key is employee number.

The first stage of normalization is to remove repeating items and to show them as separate records—but including in these separate

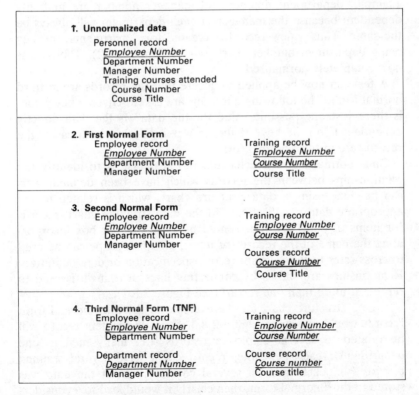

1. **Unnormalized data**

Personnel record
Employee Number
Department Number
Manager Number
Training courses attended
  Course Number
  Course Title

2. **First Normal Form**

Employee record
Employee Number
Department Number
Manager Number

Training record
Employee Number
Course Number
Course Title

3. **Second Normal Form**

Employee record
Employee Number
Department Number
Manager Number

Training record
Employee Number
Course Number

Courses record
Course Number
Course Title

4. **Third Normal Form (TNF)**

Employee record
Employee Number
Department Number

Training record
Employee Number
Course Number

Department record
Department Number
Manager Number

Course record
Course number
Course title

Fig. 2.8.2 Normalization of data (the record keys are underlined)

records the key of the original record. In part two of the example, in the first normal form there are now two records, employee record and a training record, with the key field employee number appearing in each.

The stage of going from first normal form to second normal form is to remove partial key dependencies. This is done by examining those records possessing a compound key (i.e. a key comprising more than one item) and checking to see whether each data field in the record relates to the whole key. If the field relates to only part of the key (e.g. Course Title relates only to Course Number) it is removed with its key to form another record.

The final stage of the analysis, the reduction to third normal form, involves examining each record to see whether any items are mutually dependent. If there are any, then they are removed to a separate record, leaving one of the items behind in the original record and using that as the key in the newly created record. In the example, department number and manager number are mutually dependent because the manager of each department will always be the same. Thus a new record is created for departments, the key being department number (part four of Figure 2.8.2). The data is now completely normalized.

A test can now be applied to ensure that the records are in third normal form. The following questions are asked: Given a key value, is there just one possible value for the data? Is the data directly dependent upon the key? If the answer is 'Yes' in both cases, the records are in third normal form.

Once normalization is achieved, the next task is to identify the relationships between the records which have been defined. This can be done using a data structure chart, which is related to the appropriate data specifications. In the data structure chart, records (or items of data) are represented by a rectangular box incorporating the name of the record (or item); an upper stripe can be used to cross-refer to the relevant record specification or data definition. Relationships are shown by connecting lines; a triangle is used to represent more than one record (see Figure 2.8.3).

A data structure chart for the example of the third normal form records would appear as Figure 2.8.4. Each department record will be related to several employee records (i.e. many men in one department); each personnel record could have several training records (i.e. each man on several courses), etc. If the data was stored in third normal form, then clearly it would lead to a considerable saving of storage, since manager numbers would not be dupli-

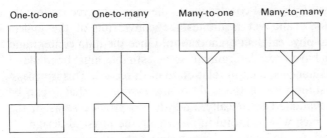

Fig. 2.8.3   Data structure relationships

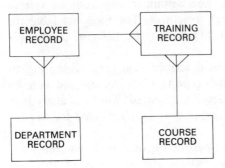

Fig. 2.8.4   Data structure chart example

cated for each employee in the same department, and course titles
would not be duplicated for each employee who has been on the
same course. On the other hand, physical relationships would have
to be established between the employee record and the department
record (to discover the employee's manager number), between the
training record and the course record (to discover the course titles)
and between the employee record and the training record (to
discover which employees in a department have been on particular
courses). Once such relationships are established, it is much easier
to process data stored in third normal form than in unnormalized
form because several keys to the data are available. If even more
keys are required, then other data items in a particular normalized
record can be designated secondary keys.

The advantages of carrying out this analysis into third normal
form are found mainly in database design where a major aim is to
build up related sets of non-redundant data. But it is also a useful
way of analysing and defining logical data requirements at the
logical system design stage.

Normalization discourages the analyst from viewing the logical data in terms of physical constraints of file sequence, record length, access methods, and yet facilitates the conversion of the logical structure into physical data organization. Once the data content and relationships have been determined, some estimate must be made of volumes, frequencies, etc., in relation to each record. This is necessary to gain some idea of the scale of the system so that it can be costed and evaluated technically, and also to provide some of the constraints which will be useful in producing the physical design.

In addition to defining and analysing data it is necessary to analyse the functions performed by the system, the forms used in the system, the flow of information within, to and from the system — and the existing organization structures. The tools available for performing these analyses are (1) flowcharts, (2) grid charts, (3) structure charts.

Flowcharts have already been described. They record the logical relationships between constituent parts of a system and can be drawn to whatever degree of detail is required. They aid analysis in that each constituent part identified may be challenged for necessity, relevance, etc.

Grid charts are two-dimensional tables; one list is entered horizontally, the other vertically. Where the elements of the lists intersect in the body of the table relationships can be expressed. Figure 2.8.5 shows a completed grid chart with the following two lists: (1) departments involved in a sales order process; (2) documents used by them. The body of the table shows the handling sequence of an order document.

The structure chart most commonly seen is the 'family tree' type for portraying an organizational structure (*see* Figure 2.8.6). Unbroken lines represent the reporting structure; dashes may indicate working relationships at the same operational level; dotted lines may indicate committee membership.

Flow diagrams and so-called string diagrams are used by O & M practitioners to indicate movements of documents between work stations and their volume. The flow diagram shows the flow sequence for one cycle of activity whereas the string diagram (it really can be string on a pinboard diagram!) shows a line per document (or multiples of documents) to give a diagrammatic picture of traffic volumes. Figures 2.8.7 and 2.8.8 show an example of each.

Statistical analyses may be useful in rendering masses of data more meaningful — for example by the calculation of averages. Comparative analyses may be undertaken by the calculation of averages and measures of dispersion (e.g. standard deviations).

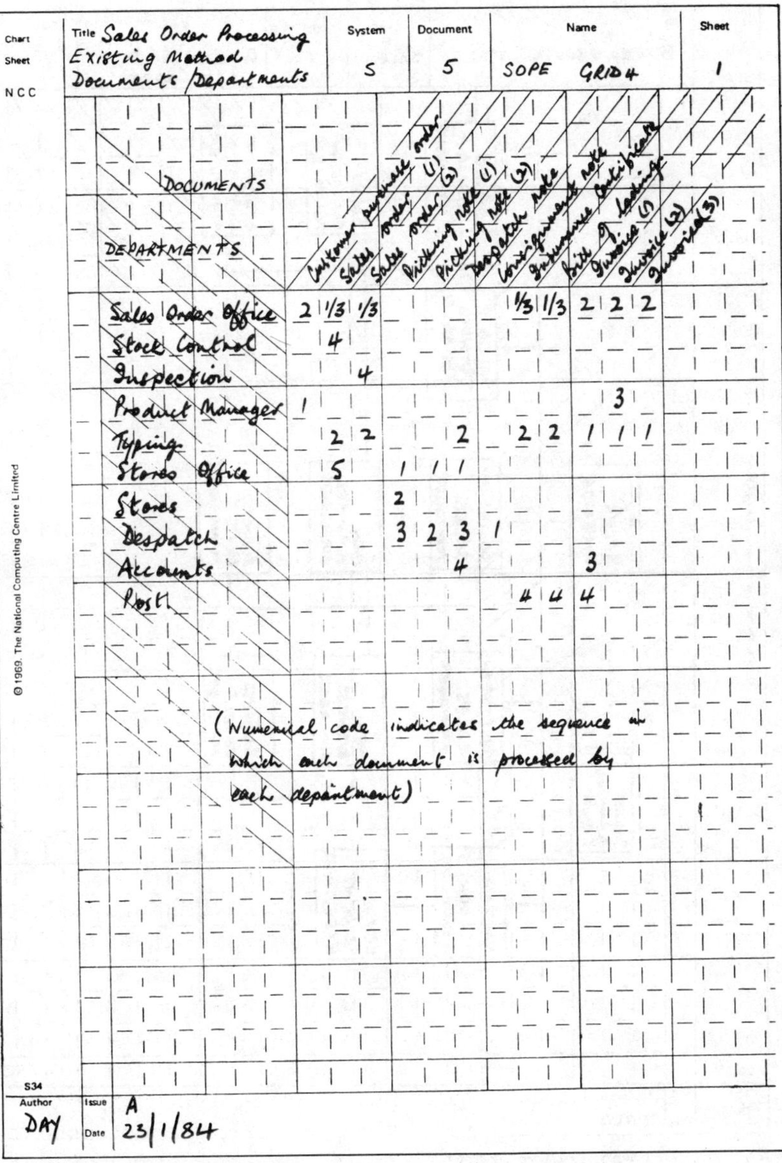

Chart Sheet NCC	Title *Sales Order Processing* *Existing Method* *Documents / Departments*	System S	Document 5	Name SOPE GRID 4	Sheet 1

DOCUMENTS

DEPARTMENTS

Document columns (diagonal labels): Customer purchase order (1), Sales order (2), Sales order (3), Picking note (4), Picking note (5), Despatch note (6), Consignment note (7), Insurance Certificate (8), Delivery loading (9), Invoice (10), Invoice (11)

Department	(1)	(2)	(3)	(4)	(5)	(6)	(7)	(8)	(9)	(10)	(11)
Sales Order Office	2	1/3	1/3				1/3	1/3	2	2	2
Stock Control	4										
Inspection		4									
Product Manager	1								3		
Typing		2	2		2		2	2	1	1	1
Stores Office		5	1	1	1						
Stores			2								
Despatch			3	2	3	1					
Accounts				4				3			
Post						4	4	4			

(Numerical code indicates the sequence in
which each document is processed by
each department)

S34 Author DAY	Issue A	Date 23/1/84

Fig. 2.8.5 Completed grid chart

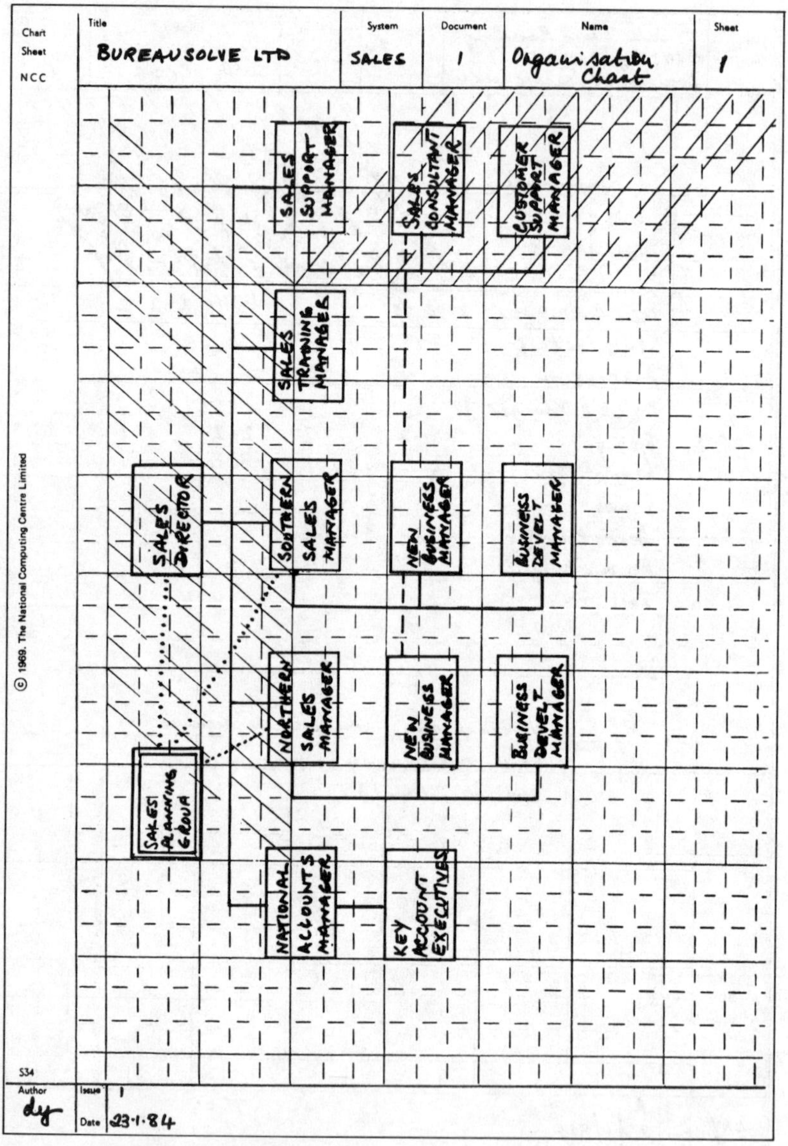

60 Basic Systems Analysis

Fig. 2.8.6 Organization chart

Trends may be analysed graphically or mathematically. Seasonal variations in volumes of data may be found. It may be important to establish the degree of correlation between two sets of data (e.g. daily temperatures and sales of ice cream) and there are well

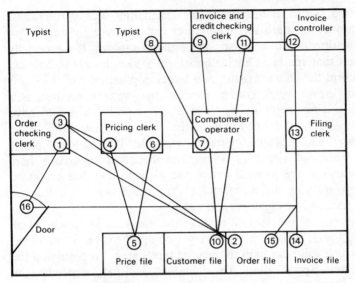

Fig. 2.8.7 Flow diagram

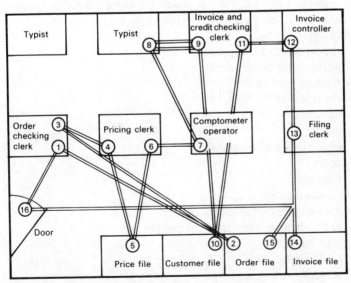

Fig. 2.8.8 String diagram

established statistical techniques for calculating such correlations. The majority of statistical techniques can be used by an analyst under the direction of a statistician or mathematician. It is especially important that specialist advice be sought if samples are to be used to represent the characteristics of a larger population.

It must never be forgotten, though, that systems analysis is a problem-solving methodology. The business problem should have been identified and defined early in the study—objectives will have been stated and quantified wherever possible. The analytical phase will be concerned with re-examining the objectives (redefining them if necessary) to see to what extent the work to date has produced ideas for satisfying these objectives. Any deficiencies revealed will call for further investigations and analysis.

At the end of the analytical stage the analysts, in collaboration with the users, will have produced a comprehensive set of documents which will later help to compile the System Proposal and the User System Specification. Their contents will vary depending on whether the system under investigation is an existing clerical system or a computer-based system; whether the organization is a first-time computer user or an experienced computer user; whether user management and staff are sophisticated or naive computer users; whether the application under investigation is complex or simple, large or small.

A full System Proposal (to be produced before detailed design work is authorized) may contain the following:

Title page
   – report title and reference
   – author and department
   – date of publication
   – distribution list
Contents list
Summary
   – objectives and proposals
   – costs
   – benefits
Recommendations
   – management decisions required
   – draft terms of reference for further work
Scope of study
   – background
   – terms of reference
   – objectives of the study

- objectives to be met by the proposal
- security and audit considerations
- time and cost constraints

Existing system
- the organization under investigation
- formal descriptions of the system (data flows, volumes of work, etc.) using charts, tables and statistical analyses
- problem areas

System requirements
- design criteria
- evaluation criteria
- future extensions

Proposed system
- outline   .
- comparative analyses of alternatives and reasons for rejection
- management implications
  - re-organization
  - computer facilities required
  - supporting services required
  - training
  - accommodation
  - staffing
  - security and audit
  - operating schedule

Development and implementation plans

Costs
- to date
- to continue project
- operating costs

Benefits
- tangible:
  - direct (staff, equipment, accommodation, etc.)
  - indirect (better resource utilization, lower stock levels, etc.)
- intangible:
  - information (quality, timeliness, etc.)
  - control

Appendices (containing details, supporting statistics, etc.)

3 Output Design

Output is a somewhat confusing term. It is used by computer personnel when referring to the transfer of data from a central processor to magnetic media such as tapes, discs and drums. It is also used to describe what emerges from a computer system for the user to act upon. This chapter is concerned with the latter.

3.1 The methods of outputting information

There are four principal methods of outputting information for people to act upon: (1) printing, either by impact printer (hammer striking ribbon and paper) or by non-impact printer (ink-jet, electrostatic, etc.); (2) screen display; (3) microfilm; (4) synthetic speech. Method (1) may be further categorized into (a) character or serial printer (one character at a time printed); (b) line printer (one line of characters printed at a time) using either a drum printer or a chain printer (these latter two types of printer are only different technical approaches to the construction of a line printer).

In methods (1) – (3) the analyst will be required to define the layout of information according to formal standards, partly in order to facilitate the programming of the requirements.

However, the ease or otherwise of programming should not be allowed—without very good reasons, such as substantial cost savings—to override the user's requirement to have information output in a way most helpful to him.

3.2 The user's role

Output design is an area of work in which the user should play a significant role—although he needs the professional assistance of a systems analyst to describe the operational characteristics of the various devices for producing output and to ensure that his requirements are practicable, within the cost constraints, and do not conflict with other system requirements.

Working jointly, the analyst and the user will have to define the following: (1) type of output, (2) content, (3) format, (4) location required, (5) frequency, (6) response time required, (7) volume statistics, (8) sequence, (9) post-printing requirements.

3.3 Types of output

The term user is ambiguous. To a computer manufacturer a computer-using organization is a user. To the computer services department of an organization anyone external to the department who receives output from the computer is a user. If there are different layers of users the ones conceptually furthest from the system are sometimes called the end-user or end-customer.

It may therefore be helpful to categorize outputs as follows:

(1) *External outputs* These go outside the organization (e.g. invoices, orders, tax returns) and therefore call for special attention as they may directly affect the organization's business relations with its customers.

(2) *Internal outputs* These remain within the organization but still require careful consideration because they may affect the operational efficiency of the total system.

(3) *Computer operations output* These are required within the computer services department (e.g. usage statistics, control reports, etc.).

3.4 General considerations to bear in mind

The Appendix, Forms Design, will be helpful to the analyst when dealing with the design of output — especially printed output.

Although line-printers are easily the most common type of output device employed, there are, in fact, numerous alternatives. The choice between them depends on such factors as the response time demanded by the system; what quality of output is required; how many extra copies are needed, if any; whether a hard copy record is desirable; what peripherals exist or can be obtained; whether the system is integrated or not; and the effective cost. In justifying the choice of output media the following questions should be posed:

(1) Is the cost justified by the benefit?

(2) Is the user's definition of requirements correct?

(3) Could two or more outputs be combined to provide a multi-use output?

(4) Is the stated frequency necessary?

(5) Is the response time shorter than necessary?

(6) Should it be on demand rather than automatic?

(7) Should there be more user control over contents and format and sequence?

The media available for providing output from the computer system are listed in Section 12.1, where input and output peripherals are discussed. The analyst should remember, however, that some central processors will allow the attachment of only specified peripheral units, and must beware of designing a system which requires the use of a particular unit, for example, only to find that this device cannot be fitted to the computer system employed by his installation. The manufacturer's technical data, supplied with the computer, will always specify the range of output media available, and this imposes a constraint on output design.

The analyst must give some thought to the presentation of information on his output documents, and to the editing facilities needed in the programs. Output editing ranges from the trivial tasks of suppressing unwanted zeros and formatting lines correctly to the complex ones of selecting records according to predetermined criteria and combining these by additional processing to provide new figures for reporting. In the latter task the standard software packages, called report generators, can often be of assistance. At a higher level still, editing can become very complex when record selection has to be performed on variable or incomplete data and the job then merges into the sphere of information retrieval.

In considering output design, the analyst must also employ feasibility checks where these are relevant to highlight suspicious events before documents are distributed. For example, in a payroll run, if a pay-slip is printed showing a negative net pay figure, this should give rise to an investigation. Finally, the clerical procedures associated with output should be considered jointly with the user and/or the O & M staff of the organization to ensure that they are acceptable to the user and are as efficient as possible.

There is one special group of forms which are more closely related to computer output — these are forms which are printed on continuous stationery. Such forms are governed by the same rules of good overall design, but a few additional points must be borne in mind.

When considering the use of continuous stationery, it is wise to make full use of the technical advice services offered by the specialist manufacturers in this field. Feeding mechanisms on high speed printers work to very fine limits of accuracy and the stationery must be printed with similar precision. It is necessary to remember that there are additional costs involved with this stationery over and above the printing itself, because of the paper-handling equipment required.

It may be necessary subsequently to guillotine printout. It is surprising how often, after the printing run, it is discovered that no suitable guillotine is available, or, perhaps, even exists at all! It is important to check that equipment is available to do this operation when planning the job and to include the guillotine costs. It is also worth remembering that while multiple printing may fractionally save time, the printout may not be usable until further time is spent in guillotining. Even though guillotining may be an off-line job, it may add several hours to the completion of a tight schedule job such as payroll. As an aid, it will be beneficial to include a mark at the guillotining position so that a constant check can be made. Guillotining is usually done simultaneously with decollation and not on a gang basis. Consider what size the printout will be after guillotining. If it is very narrow it will be difficult to handle without twisting or tearing. Moreover, the problems of filing which apply to all small documents are significant. Folders to carry such printout singly are not available as standard and will be costly to provide.

Where the total character width on the printer is a multiple of the maximum number of possible characters in the form, it is often considered desirable that the printout be printed two or more up across the paper.

Much depends upon the destination of the printout and the degree of user involvement in the design of the output form. Two-up printing may have many advantages but there may be processing difficulties. The necessity to assemble two or more sets of data in storage, and problems with overflows from one form to the next, may cause programming problems. If decollation is required for the printout and the decollator will simultaneously guillotine, the saving in paper and time may well justify the use of such a printing technique. Where guillotining has to be done separately there is generally greater merit in printing singly. Ease of programming and printer speed are not necessarily significant but can be so in certain cases. It should also be noted that on modern computers printing is often semi-off-line or completely off-line. Unless the installation is printer limited there is no saving of overall processing time in using complex printouts. Effective printing speed is the guiding factor.

In the draft and final design stages of the form which is to be output, the following format and handling considerations apply. The overall form width should be kept to standard stock sizes (1) to reduce paper costs; (2) to reduce set-up time in form-feed devices; (3) to reduce costs of filing and binding; (4) to reduce the variety of binders in use.

Any vertical lines should split a print position otherwise registration is difficult, or impossible, due to paper shrinkage and slight machine variations.

Line spacing on printers is generally either six or eight lines per inch. Sometimes double spacing at eight lines to the inch could be considered instead of double at six, as this gives 25 per cent more lines. The number of preprinted lines should be kept to a minimum as an aid to legibility. It is often very useful to make a sample study of the frequency distribution of the occurrence of body lines, e.g. items per invoice, before designing the form and so save space and also reduce the number of long skips or overflows to the next form. Skipping must be minimized, although this is obviously associated with the lines occupied by variable data. The total number of body lines should be a multiple of the spacing used. Some lines depend on calculations before they can be printed, and they must be placed on the form so that printing is not delayed by waiting for calculation.

It helps if a form alignment guide mark, such as a small cross or dot can be preprinted. If it is not present the operator will find some way of determining when the form is set up properly, but this usually complicates the set-up instructions and may lead to waste.

When marginal perforations are used, allow about 0.5 in between the perforation and the first printed character. The practice of using the form feeder holes for binding is increasing, particularly for internal documents. Usually, both sides of the form are left untrimmed.

Subject to cost, form titles and other headings should be preprinted wherever possible as this is neater and uses less computer printing time. For internal documents where preprinted headings are not desirable or possible, the use of a transparent template by the user may be considered.

Impact printers can print on multi-part stationery with sufficient power and clarity to produce acceptable output on six-part forms.

Speeds, however, are limited to a maximum of about 3,000 lines a minute. Newer, non-impact laser printers, while unable to cope with multi-part stationery, have printing speeds up to 20 000 lines a minute and give excellent print quality. In theory therefore, six-quality can be produced in less time. Additionally, no decollation is necessary and guillotining can be done by a printer attachment. Furthermore, different type founts and graphics can all be incorporated on a single page without any loss of speed, and colour output further enhances the appearance of the output. Laser printers are expensive of course, but with costs falling are now much easier to justify on the grounds of cost.

Similar facilities can be obtained at less cost using ink jet printers but speeds are much lower. These micro-processor controlled devices shoot dots of ink through a perforated metal plate and are very suitable for use in the automated office where high quality output and quiet running are required.

3.5 Output definition

As the user and analyst progress on the design of output forms, displays, etc., it is desirable to record the decisions taken in order to record unambiguously the stage they have reached, to help in communicating their ideas to other users and to the computer services department.

The NCC Data Processing Documentation Standards provide forms for this purpose. Headings, subheadings, some representative data items and totals may be entered on the form, character by character, to give a representation of what the printed output will look like.

In addition to the layout representation there must be a formal written description of the data contents of the form. Each type of form to be printed may be regarded as a computer record-type and so the NCC Record Specification form may be used for this purpose —*see* Figure 3.5.1 for example. This specification is essential for the programmer.

Finally it is necessary to ensure that all the essential characteristics about the document are formally recorded, and the NCC Computer Document Specification form (Figure 3.5.2) is a good example of what is required. The conventions used to describe the logical structure are those used to describe computer files. All the

Record Specification NCC — Record description: *Order Acknowledgement Details*

System	MSOP
Document	4·6
Name	ACK.ITEM
Sheet	1
Lay-out chart ref.	ORDACKPL5
File specification refs.	ORDACK
Record size	6400 Max
Record format	Fixed ☑ / Variable ☐
Record length	Fixed ☑
Medium	PRINTER

© 1969. The National Computing Centre Limited

Ref.	Position From	To	Level	Name (In system design / In program)	Data Type	Size	Algn mark	Picture	Occurrence	Value Range
*	Maximum of 28 lines per page of items									
1	>		01	ITEMLINE	C	80			1–80	
*	1st line of items in line 23									
2	12	13	02	SERNO	C	2		99		01–99
*	Serial numbers are consecutive over the pages of a single acknowledgement									
3	17	20	02	PRODCODE	C	3		A99		A–Z excl I,O,U / 01–99
4	22	46	02	PRODDESC	C	25		X(25)		
5	48	51	02	QTYORD	C	4		ZZZ9		1–999

S 44
Author DAV
Issue A
Date 1/12/84

Fig. 3.5.1 Record specification for a printed report

Computer Document Specification	Document title	System	Document	Name	Sheet
N C C	Sales Analysis (Area)	SA	4.6	SA(A)	1

Stationery ref.	Width	Depth	Number of parts
Standard listing			2 Blank/pre-printed

	Average	Maximum	Growth rate
Pages	ten per area total 40	12 60	nil % per or determining factor
Lines per page	16	16	

Page and line spacing/stepping double line spaced

Ribbon type	Ribbon life	Printer speed	Lay-out chart ref.	Control loop
normal			SA(A) 1	Ref. normal

OUTPUT ONLY

Part No	Trim/Burst	Distribution	Line	Channel
1		Area sales manager		
2		Sales director		

MACHINE READABLE ONLY — Clear area – distance from edges | Reading method and font | Source

Level	Record Name	Size	Unit	Format	Occurrence
GA	Sales Analysis by Area				4
GB	Account Analysis	2 – 38	L		1–12
C	SA(A) Headings	3	L		
C	SA(A) Details	1	L		

S 43 — Author — Issue A — Date 8/10/84

© 1969 The National Computing Centre Limited

Fig. 3.5.2 Computer document specification

records relating to a particular computer output constitute a file; each record-type is a form-type.

When output is destined for a display screen, the NCC Display Chart is helpful in designing the screen layout of information passed to and from the computer. The chart is used to depict the format of a complete message to be input or output. Some guidelines for the design of dialogues between the terminal user and the computer are as follows:

(1) They must be easy to learn and use.

(2) Guidance to the user must be available on the screen, if required.

(3) The dialogues should use the user's terminology.

(4) Response times should be as expected and needed by the user—otherwise he should be told that there is a temporary delay.

3.6 Computer output on microfilm (COM) and voice response units

These are less likely to be encountered by the junior analyst than printed output on stationery or displayed output on screens. Nevertheless COM has been available for many years; it is especially useful for archiving large volumes of information, where there is an occasional need to retrieve information speedily in human-readable form; or for disseminating large volumes of information to dispersed sites for answering *ad hoc* enquiries, etc. It is used in a wide range of applications, particularly for stock lists, parts lists, directories, payroll records, sales invoices/statements, accounting records, and financial reports.

Microfilm is rarely used now. The standard COM output media is microfiche: sheets of film approximately 6 in. by 4 in. Each sheet of microfiche can contain up to 270 pages of computer printout with a title and index number which is legible to the eye. COM is usually produced by an offline recording device which receives its input from a magnetic tape; this tape will have had print-line images written to it by the computer system. The film, or fiche, is subsequently read by using a film or fiche reader (or viewer) to magnify the recorded information and display it on a screen. There are various methods of indexing to facilitate easy retrieval. The same general principles of form design apply to microfilmed information as to printed information.

Voice response is still very much in its infancy but undoubtedly will become a common form of input and output to computer systems in the near future. It is already in use in various process control applications and for some banking terminals. One obvious advantage is that it is potentially a much more efficient method of input than a keyboard. The response is either generated from stored sounds or synthesized from coded information. The former results in a distinctly human-sounding response whereas the latter is machine-like. However, rapid progress is being made in improving the quality of synthetic speech.

4 Input Design

Like output, input is a confusing term because to computer people it often refers to anything being transferred into the central processor, e.g. from magnetic discs. However, this chapter is concerned with data originating outside the computer system.

4.1 The method of inputting data

There is no doubt that data capture in its totality (from data origination to recording on a computer storage medium) is costly. For business systems it may well account for 25 per cent of the total costs of a system over its entire life. In many cases data input is labour intensive and therefore it is time well spent designing forms which are speedily and accurately completed and capable of being efficiently recorded on computer storage medium. Thus the designer has these objectives in mind: (1) to produce a cost-effective method of input; (2) to achieve a high level of accuracy; (3) to ensure that the input is acceptable to the user.

There are four ways in which source data can be input to a computer system:

(1) It can be converted into a computer-readable medium by a person using a keyboard. The commonest media are cards, paper tape, magnetic tape.

(2) It can be read directly by a computer peripheral device. The commonest examples are as follows: optical mark reader (OMR), optical character reader (OCR), magnetic ink character reader (MICR), magnetizable plastic card readers and punched tag readers.

(3) It can be keyed directly into a computer from a remote terminal—such as one finds in a multi-access, on-line system.

(4) It can be produced as a by-product (e.g. of an accounting machine) and read directly into a computer (e.g. magnetic tape cassette reader). The choice of method will depend upon satisfying

the objectives already stated. If the user requires a response to an input message within seconds, then an on-line terminal such as a visual display unit (VDU) is almost certain to be the correct method and design of input is likely to be concerned with screen layout and dialogue design. In other instances the source data arises in such a way that it would be impracticable to read it using OMR, OCR or MICR—because MICR requires a special print format and mag-netizable ink, OCR may require a special print fount—but certainly will require good quality printing or hand-blocked characters, OMR may be too restrictive in that only marks can be read on a docu-ment. As a result a vast amount of data has to be converted by human operators—reading each character of source data one by one and striking the appropriate key on a keyboard; the conversion medium may be cards, paper tape or magnetic tape.

4.2 The input process

Whichever method is chosen it will include some or all of the following: (1) data recording at source; (2) data transcription on to an input form; (3) data conversion to a computer-readable medium; (4) verification of data conversion; (5) data transport to the com-puter; (6) data validation and authentication; (7) data correction; (8) control at all stages over the movement and processing of data.

4.3 The user's role

As with output design, input design is an area of activity in which the user should play a significant role. Ideally the user should be responsible for the accuracy and timeliness of data being input to a computer system—the quality and timeliness of the output informa-tion is almost wholly dependent upon the data input (remembering garbage in, garbage out!). It is imperative, therefore, that the systems analyst should act only as guide and adviser in this area; to impose something against the user's wishes could well be counter-productive.

These are the aspects the user management ought to be con-cerned with: (1) costs; (2) staffing, training and supervision; (3) productivity; (4) quality control; (5) accommodation and environ-ment.

Here are some questions the analyst should seek answers to when considering input design and its relevance to the users who will be originating the source data:

(1) Are the originators employees or customers?
(2) What training will they need?
(3) What is their motivation?
(4) How will they react to change?
(5) In what kind of environment do the originators work?
(6) Are there constraints on the devices they could use?
(7) What response time is appropriate for the data?
(8) What kind of service is required: validation? interrogation? up-dating?

4.4 Types of input

It is helpful to bear in mind the type of input the analyst is concerned with designing. They may be categorized as follows: (1) external to the organization, e.g. purchase invoices; (2) internal communications originating within the organization, e.g. new customer record; (3) computer operations input, e.g. job control parameters.

Some of these inputs may be on-line, using VDUs in interactive mode; others may make use of turnround documents which have been prepared by the computer in an earlier process and are later input as a transaction to update a master-file record (e.g. a bill can be output on a computer line-printer, despatched to a customer, returned with a cheque and input to the computer via OCR); others may make use of specially designed documents for data capture which have to be converted by a keying operation into a computer-readable medium.

4.5 General considerations to bear in mind

Inputs to the computer system can be either batched records or real-time accesses via some kind of terminal. Even with the latter method of data acquisition it may still be desirable to use batch processing, and this can be achieved by the use of a terminal which will transcribe the data into a suitable medium such as punched paper tape. This can then be processed in batches at a convenient time later in the system. Regardless of the method of capture, the analyst must consider if his input is relevant to more than one system and design it accordingly. An example of this could be an order from a customer which might have an effect upon invoicing, sales analysis, stock control, production scheduling and transport systems, each of which would require different information. This may mean that the transaction should carry more detail than is required for any one of its several uses, or that it should be related by reference to a previous transaction. In turn, this often requires the analyst to consider if the transaction should be preprocessed and then distributed, or distributed in its primitive form; and more important still, it leads him to a consideration of how far integration should go. Integration is essentially the technique of making multiple use of input data and of using the output of one subsystem as the input of another.

It is normally true that the more integrated a system becomes, the more efficient is the use of the computer. However, the systems design needed for an integrated system is far more complex than for one which treats each subsystem in its own right and a compromise often adopted is to plan an integrated system but to write each subsystem with its own unique input/output. When all the component systems have been successfully implemented, the input and output sections are modified and integration achieved by stages. This approach gives users of the system more confidence than an ambitious integrated system which will not hang together in operation.

When batch processing methods are being used it is normal to insist that file maintenance data be submitted as a separate batch, so that it can be handled before current transactions, thus ensuring that the file is as up-to-date as possible. Sometimes, run considerations in an installation determine that file maintenance runs should be performed at intervals, say weekly. Following the maxim mentioned above in these cases it may be safer to have amendments, etc., submitted each day and to store them at the head of the main file until the regular updating run is processed. In this way, the possibility of error through non-preparation and loss of documents is minimized.

When determining the inputs to a system, the analyst must recognize the characteristics which form the substance of an input item.

The specification for each type of input to the system must detail the following: (1) identification—for recognition, including batch headers, etc., if relevant; (2) content and format—in specific terms, including media used; (3) frequency of receipt; (4) expected volumes—in suitable units, i.e. documents, characters, etc.; (5) conditions which govern its appearance in the system; (6) sequence in which it will be received for processing; (7) validation procedures to be used to ensure accuracy.

The analyst must also decide on the input media to be used in the system. A number of points must be considered, including the basic input philosophy (real-time or batch processing), what records of input are required, what peripherals exist already or can be obtained, whether the system is integrated or discrete, and last but not least the effective costs.

In suitable circumstances interactive, 'conversational' input procedures are possible using keyboards working directly into real-time systems, and systems of this kind can be expected to proliferate. The design of the system's responses is a complex task and usually involves a great deal of development work to ensure that the system is pleasant to work with and that the responses are always positive and meaningful.

Clerical input procedures are a subject in their own right and the well developed techniques of O & M or clerical work-study should be applied to ensure that clerical procedures are efficient. The design of the computer input system represents a constraint on the design of the clerical system and it is important that the existence of this interface be realized and that neither the computer nor the clerical system be designed in isolation.

4.6 Data capture

There is a need to reconcile possible input methods with the operating requirements of the user. It is of little use to expect a salesman to complete a complicated form in a customer's office, or to expect factory workers to comply with sophisticated and complicated methods of production control. Also, the provisions of input may be the only contact that the majority of employees in a company have with the computer department. If they regard this provision of input as a tedious and unrewarding activity, they will be resistant to data processing activities and the quality of information provided will suffer. It follows from this that simplicity and ease of data capture and collection must be reconciled with the need to present data to the computer in a format and sequence that is acceptable.

Ideally, the data should be recorded in a machine-processable form as a by-product of the actual transaction or event as it happens. This is possible, but can be both expensive and unsatisfactory for other reasons. Economy, availability of trained staff, machines and servicing capabilities frequently cause centralized data preparation. If centralization is to be adopted, then consideration must be given to the transmission of data. Most methods are suitable for operation on either a centralized or remote basis.

The methods of capturing data are numerous. The choice of a particular method among them depends largely upon the application involved, the overall system timing requirement, the volumes of data to be processed, and the cost of the equipment in relation to the benefits derived from its use. The equipment available is as follows:

(1) *Key-to-disc systems*   A mini- or microcomputer controls a number of keying stations, each of which is equipped with a VDU. The computer typically has disc storage, a magnetic tape unit and a printer. As input records are keyed in at the work stations they are accumulated on disc, checked and validated. Batches of input records are totalled and checked before being transferred to the magnetic tape, from which the input is in turn transferred to the main processor. The printer can be used to produce a hard copy of the input and also to output operational statistics. The amount of checking and validation which is performed prior to transferring the data to the main processor reduces the amount of data validation needed on the main processor and ensures that most errors are corrected at the point of initial input.

(2) *On-line terminals*   These will typically consist of a VDU and keyboard, but there may be a printer/keyboard device, a microcomputer, or even a mini-computer with VDU, keyboard, printer and batch input facilities. If the terminal is in the same location as the main processor it can be linked directly by cable, but remote devices will have to be linked via a data communications network. Input data may be transferred in batches to the main processor for later processing or individual transactions may be keyed in which are verified automatically before processing.

(3) *Character recognition*—by a range of devices. Some are referred to as 'page' readers, others 'document' readers; the distinction is not clear. Generally a 'document' is a piece of paper with a relatively small amount of data on it whereas a 'page' is large (e.g. A4 size) and contains a relatively large amount of data to be read.

Page readers generally have medium speed paper transports coupled to sophisticated scanning devices whereas document readers have simpler reading devices but concentrate on high speed paper transportation.

Optical mark readers (OMR) The page-reading devices read hand- or machine-made marks on specially designed forms. The converted data may be output off-line on to any desired input medium or input on-line directly to the main processor (either locally or via data transmission facilities). They are more suitable for numeric than extensive alphabetic data. They eliminate the need to convert the data by keying. Paper feed rate is 300–3000 sheets per hour. The efficiency of this approach depends on how adept the originators are in entering numbers as marks on a form.

Optical character readers (OCR) Page-reading OCR devices read not only marks (as for OMR devices) but also hand-printed characters, typed characters, computer-printed output (e.g. forming turn-round documents). The most sophisticated OCR devices will read not only full alpha-numeric character sets but multi-founts. Throughput is very much dependent upon the packing density of the data on the page—but the limiting speed of the scanner may be about 2000 characters per second. Document-reading OCR devices (e.g. for vouchers, counterfoils, receipts, etc.) concentrate on higher paper transportation and can cope with, say, 35 000 documents per hour.

Magnetic ink character readers (MICR) These document-reading devices read highly stylized founts printed in magnetizable ink. Badly crumpled or defaced documents may be read satisfactorily. Documents are transported at up to 45 000 documents per hour—and sorted into output stackers, if required. These devices are confined almost entirely to reading cheques and similar documents.

(4) *Card punches—unbuffered* These are electromechanical devices for automatically producing punched cards by a keyboard operation. Most commonly they use the 80 column card. About 8500 key depressions per hour can be regarded as an average work rate.

(5) *Card punches—buffered* Similar to the other type of punch but with a built-in dual buffer memory; memory is also used to hold format, correction and control programs. The buffers work alternatively—while the operator keys data into one buffer the contents

of the other are being punched. Skipping and duplication (ganging) take place at electronic speed. In the verification mode a card is read into a buffer and compared electronically with the key-strokes; incorrect cards are corrected in the buffer and repunching done while another card is being verified. Overall throughput may be 25 per cent higher than for an unbuffered punch. Additional facilities include check-digit verification and operational statistics. The use of punched cards as an input medium is declining rapidly as key-to-disc and on-line terminals provide for greater efficiency.

(6) *Paper tape punches* These facilitate the punching of variable length fields and records. Work rate may be 25 per cent higher than for unbuffered card punches. Data on paper tape is difficult to amend; spliced tape may be troublesome to read by paper tape reader. Paper tape is rarely used these days, although it can facilitate data transfer by telex machines, thus not requiring modems. However, the speed and limited character set offered by telex still do not make it an efficient method of transfer.

(7) *Point of origin devices* These fall into two main categories: factory or shop floor; point of sale.

Factory or shop floor devices are able to read a combination of fixed data (e.g. from punched cards, plastic badges, etc.) and variable data (e.g. from a keyboard). Such devices may have a screen display—useful for tutorial guidance. They may be off-line (producing cards, magnetic tape, floppy disc, tape cassette, etc.) or on-line to a main processor. Some devices use simple hand-held OMR devices for reading printed bar-codes and converting them into computer codes on magnetic tape cassettes.

Point of sale (POS) devices are effectively cash registers with a recording or transmission facility. A simple POS device merely records data on a computer-readable medium (paper tape, magnetic tape, etc.). A complex POS device may well be part of a real-time computer system and perform local checking and data manipulation; it may offer the ability to read tags, labels, badges and credit cards.

Point of sale tags may be Kimball tags (commonly used in the retail clothing trade) which are converted by batch readers into paper tape or magnetic tape—or magnetic tags which may be read by some POS devices or by portable readers (for stock-taking).

(8) *By-product systems* The following devices can be equipped to produce computer-readable data produced automatically as a by-product of the principal operation for which the device is used: accounting machines; visible record computers; tally-roll add-listers;

typewriters. The output medium may be paper tape, magnetic tape, OCR-readable print, discs, punched cards.

Systems analysts must examine these different types of equipment to determine which may be the most appropriate in a given environment, and, if involved in the selection of equipment, decide which is the most suitable and economic in the light of the operating requirements of the company and the prices in force at the time. It may not be valid to reject any possible method of input on the grounds that it does not appear superficially conventional or economic. For example, in certain circumstances OCR may replace profitably a section of four punches and three verifiers.

There is a set of factors which is common to all forms of input. It is by an evaluation of these factors that the systems analyst can determine the suitability of a particular method for a particular application. These factors are the type of processing, speed, accuracy, verification, rejection rate, operator requirements, cost and relevance.

The type of processing may be batch (groups of records) or demand (individual and distinct records, processed as such), serial (sorted) or random (unsorted). For example, punched paper tape may be unsuitable for serial processing in that it cannot be sorted off-line if required. Punched cards are generally unsuitable for demand processing, as the preparation time is considerable.

Speed must be considered in capture, in preparation, and in entry to the computer. Punched cards and paper tape are slow to prepare, but relatively fast in entry to the computer. OCR documents need little preparation, but can only be read relatively slowly.

Accuracy means the true representation of original information. If transcription can be avoided, then input may be more accurate, according to the device used. If a large volume of data has to be accumulated, say for a statistical report, then a certain level of inaccuracy may be acceptable with regard to the economies achieved.

When assessing verification consider the various possibilities: they include sight checking from hard copy (poor), arithmetic methods (by check-digit verification, for example), re-entry on punched cards, or automatically by program, and finally batch-level verification.

The rejection rate will be influenced by the general unsuitability of some documents for automatic handling, while operator requirement will be determined by an appraisal of the need for intervention or total involvement of trained staff.

Cost considerations take recognition of the capital cost of equipment, as well as the cost per character of input. In the latter must be included stationery (or media), floor-space, staff and overheads. Relevance, the last in the list, involves considering the inherent suitability of a device for a specific application, e.g. Kimball tags hang well on clothes, MICR suits cheques well. Other factors which may require to be taken into account sometimes are the relevance of optional devices, possibilities for expansion and modification, and suitability for other applications.

A useful check-list to bear in mind when considering data capture in shown as follows:

(1) *Type of input* Will input be batch or demand, serial or random?

(2) *Flexibility of format* Fixed/variable length records. Does fixed length mean a high redundancy?

(3) *Speed* What are the speeds of capture, preparation and entry to the computer?

(4) *Accuracy* True representation or original data; level of accuracy demanded; transcription; method used to detect errors (sight, re-entry, parity, etc.).

(5) *Verification* Is it necessary? Where? Sight-checking. Arithmetical checking by use of check digits. Re-entry (keyboard). Control (batch-level).

(6) *Rejection rate* Does the input media permit automatic handling of errors?

(7) *Error rate* Is the error rate likely to be high or low? What effect will undetected errors have?

(8) *Correction* At what point will corrections be made? Is re-punching necessary?

(9) *Off-line facilities* Are sorting facilities required? Can the input be physically filed? Are dual-purpose facilities required? Can some part of the input be prepared in advance?

(10) *Need for specialized documentation* This is influenced by input method but consideration should be given to directly processable input.

(11) *Standardization* Ensure uniformity of coding throughout the system:

(12) *Storage and handling* Consider convenience; cost; volume; retrieval; wear and tear.

(13) *Automatic features* Consider the following features of data capture devices: check-digit verification; control totalling; parity checking; gang-punching; reproducing.

(14) *Hard copy* Is the primary objective of the job to produce hard copy? If hard copy is required, can it be produced 'off line'?

(15) *Security* Consider the following security aspects: loss; distortion of sequence; corruption of data; fraud; spoiling (dirt, damage).

(16) *Requirement for trained staff* What clerical skills will be required? Is greater accuracy required?

(17) *Relevance to particular application* Are there any special techniques which are widely used in similar application, e.g. Kimball tags (clothing industry), plastic cards (banking)?

(18) *Costs* Consider all the costs: equipment; floor-space; operators; supervision; supplies; back-up facilities; telecommunications etc.

4.7 Input specification

The documents designed for data input should be specified formally as part of the installation's standard documentation. The NCC Clerical Document Specification form is for this purpose (Figure 2.7.13). For defining the input in computer media the NCC Computer File Specification and Record Specification forms are available (Figure 4.7.1). When specifying the data the analyst must be precise in stating all details of the input. Each input for data preparation is given an unambiguous name, followed by a description of the character code used if more than one code has been specified in the general description. For input needing translation into computer media, e.g. original documents needing punching into paper tape or cards, the specification includes the following: (1) a sample of the document (in an appendix) showing whether it is an existing document or a document specially designed for punching; (2) the source of the document, e.g. time-sheets from timekeeper's office; customer orders from sales department; (3) a statement of any manual scrutiny, editing, transposition, extension, etc., which occurs before the document reaches the data processing centre; (4) a statement of any pre-sorting, batching, batch-numbering, or pre-casting for control purposes; (5) a sample of any control slips produced.

The specification briefly describes the data preparation and verification system to be used, e.g. two stage punch and verify, punch only, read back and edit.

When the input is ready for entry to the computer, either directly

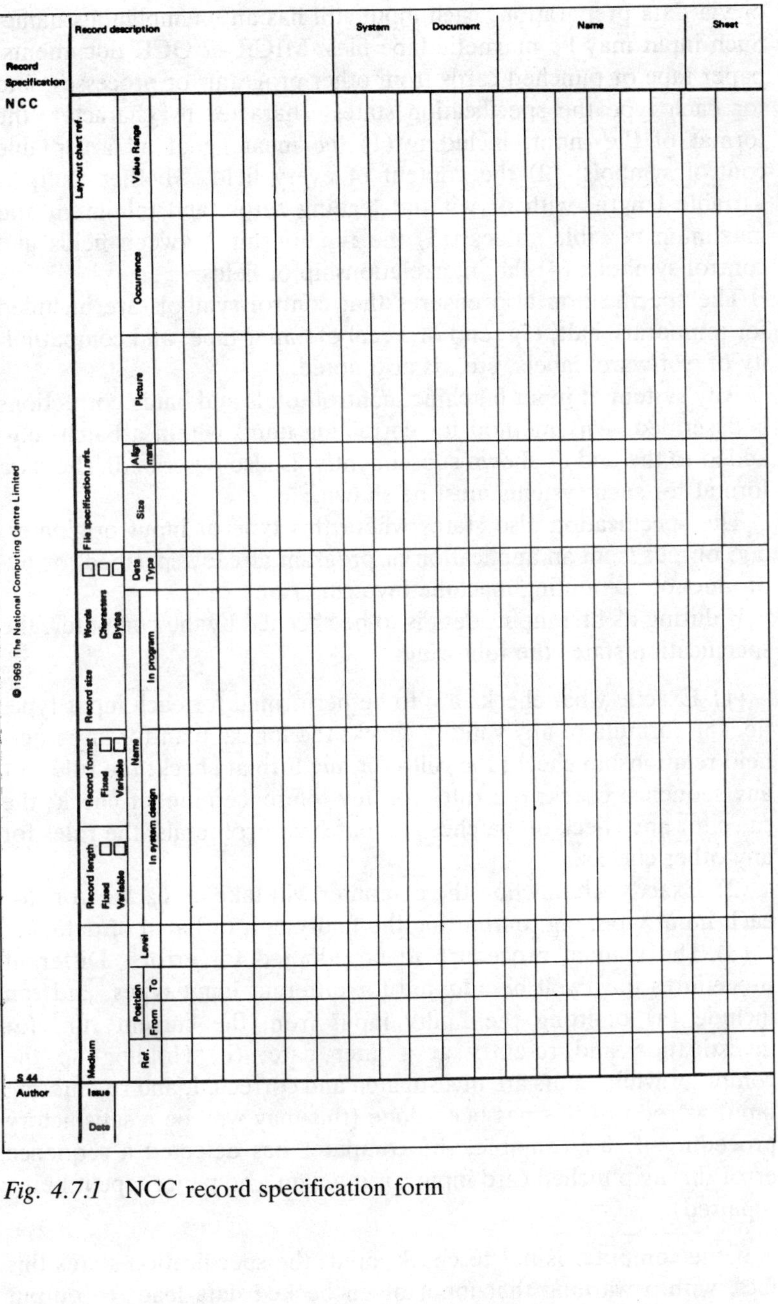

Fig. 4.7.1 NCC record specification form

or via data preparation, each input still has an unambiguous name. Such input may be magnetic tape files, MICR or OCR documents, paper tape or punched cards from other programs or processes, and for each type the specification states, character by character, the format of the input, including (1) the meaning of all acceptable control symbols; (2) the content of every field, whether fixed or variable length, with or without leading zeros, and minimum and maximum possible values; (3) the relationship between fields and control symbols; (4) the interrelationship of fields.

The specification also ensures that control symbols are included for temporary halt, e.g. end of a reel of paper tape, and compatibility of 'software' labels, etc., is also noted.

Any system of input batching, control totals and batch corrections is described. Any method for correcting items within a batch, e.g. added to the end or shown subsequently, is also described. The data format for such systems must be shown.

The specification also states where this type of input originates, e.g. output from an application or program already specified, or by-product of accounting machine invoicing run.

If during its first input, data is to be checked by the computer, the specification states the following:

(1) Exactly what checks are to be performed for each input type, i.e. the formula of any validity check; the logical plan for any inter-field relationship check; the rules for any format check; the rules for any sequence check; the rules for any reference number check; the rules for any check on batches and batch control totals; the rules for any other checks.

(2) Exactly what action the computer will take on each error, for each input type, e.g. outputting the faulty unit or batch of data.

(3) The manual procedure to be adopted for errors. Different procedures may well be adopted for different input types, and can include (a) omitting the faulty input from the current run, for investigation and re-entry at a later date; (b) Holding up the computer while faults are investigated and corrected, and re-entering input as soon as this has been done (this may well be a satisfactory procedure if, for example, the computer has detected a sequence error during punched card input; or if a damaged paper tape if being repaired).

If the computer is not to check input, the specification states this fact, with a warning that input of unchecked data leads to output results of questionable validity.

If punched cards are to be used for input, the specification will include a specimen card design stating the contents of each field. This is sometimes done on a blank card, although this is often difficult with narrow column headings. Another alternative is to use the special card drafting sheets supplied by the manufacturers which have a blank outline card drawn very much oversize and indicate the card columns by preprinted numbers at equal spacing.

4.8 Forms design

In designing clerical documents for computer input the analyst should pay heed to the general principles of forms design — see the Appendix. The efficiency of the input process may depend very much on the quality and suitability of the forms specially designed for this purpose.

4.9 Code design

It is comparatively rarely that a systems analyst will be required to derive a code of any size from first principles. Codes tend to be a hallowed part of existing systems, and to have grown with those systems. The need for designing a major code tends to arise (1) when a new system is superimposed in an existing organization; (2) when a code is outgrown and cannot sensibly be extended, because to do so would involve making nonsense of its structure; (3) when two organizations with different coding systems merge, and a single code is needed for both.

The function of a code is to make it possible to identify, or retrieve, the coded items as efficiently as possible. This applies to normal data processing activities as well as to information retrieval systems. In the latter it is the information structure rather than the item listing which is coded. In all these applications, the code fulfils this function by providing a substitute for the normal item name, which from an information point of view consists of a set of irrelevant characters. This substitute name can do two things. It can reveal the relationship with other items of a similar kind, and it can reveal properties of the item itself. A normal name does neither of these things.

A data processing code must fulfil the following functions if it is to be successful:

(1) It must be logically tailored to the system.

(2) It must be precise so that any item is described by only one code word.

(3) It must be sparing in its use of characters.

(4) It must provide for all the system expansion and development that can be foreseen.

(5) It must be clear in structure, so that the user can understand it and encode items without error. Possible sources of ambiguity must be clearly documented.

(6) Mnemonic aids should be used if clerical or other non-computer reference is to be made of it.

(7) It should suit the data processing software and hardware, the system of storage levels and the indexing systems with which it is to be used.

Systems considerations in designing a code require answers to the following questions:

(1) What are the data to be used for and in what sequence?

(2) What function does the code play in this? Who will use it?

(3) What is the structure of the data and the file management system?

(4) Are we coding data lists, or an information structure?

(5) What is the size of the file with respect to the hardware provided?

(6) What is the rate, and character, of item searching?

(7) What is the rate of growth of the file. If the data are hierarchically organized, is the tendency for the main constructs or for the low level subsets to increase?

(8) Are there any occasions on which the code will be used visually as a means of clerical identification? If so, which parts of the code word are subject to this?

Allied closely to coding is the need for classification. This can be defined as the systematic arrangement of all items within a system, so that like items are grouped together. Within the resulting framework, individual members of a group are further defined according to their fundamental attributes. An acceptable coding system will emerge naturally from well classified data. When a coding system proves to be inflexible to minor changes, in most cases the basic classification will be found to be deficient. A good example of this is to be seen in sequence codes, where the next available number from a list is allocated to each item in turn. For example

> Code no. 01 = Adams J.
> Code no. 02 = Baldock T.
> Code no. 03 = Brown P., etc.

These simple codes are rarely used for representing more than 20 to 30 items, because of their inflexibility. It is impossible to insert new items into the sequence, and the removal of old items makes for redundancy. The code contains no useful information about the items and there is no correlation between the code and the items represented.

An improvement on the sequence code is the group classification code in which all, or some, of the digits in the code number indicate a particular classification. For example

> 5xxx   Purchases
> 51xx   Production material purchases
> 511x   Steel purchases
> 5111   Steel plates
> 5112   Steel strips
> 5113   Steel wire, etc.

Note that the last digit in this example is a sequence code.

It follows that when a numerical code is used then up to 10 classifications are possible for each digit in the code, and each classification is represented by one of the digits 0 to 9. In some codes one or more digits are not taken up when the system is designed, but are held in reserve for new classifications that may arise.

When an alphabetical code is used, then up to 26 classifications are possible for each position in the code. In practice, because of possible misinterpretation, it is usual to use only 21 letters, and the code is thus restricted to 21 classifications.

The attributes of items which are to be coded do not, of course, always fall into 10 main groups, 10 subgroups, 10 sub-subgroups and so on. Where there are less than 10 groups to be classified then the code is not fully utilized, and consequently an alternative to be considered is a block code.

The block code is virtually the group classification code modified to provide more groups with less digits, but leaving room for a limited amount of expansion. For example

There are 4 main groups, i.e.	The number of subgroups in each corresponding main group is
A	15
B	21
C	18
D	24
Total number of groups	78

The normal group classification would require a three digit code for these 78 items. One possible solution using a two digit block code is

Main groups	Code numbers allocated	Spare code numbers in each group
A	01–19	4
B	20–46	6
C	47–69	5
D	70–99	6
		21

Block coding systems of this kind are common and easy to expand within each category. They can, with practice, be associated with the items or groups they represent. Whole sections can be added, or deleted, and the user or computer can carry out simple checks on the numbers as they are received. The code can often be used as a basis for sorting and limited information retrieval.

Another simple coding technique is the use of significant digits. In these codes individual digits are made to represent features of the coded item, or are given some other special significance. In general, the term refers to codes containing actual facts about items. For example

TT 670 15 B = Tube type, size 670 × 15, Blackwall
TT 710 15 W = Tube type, size 710 × 15, Whitewall

The use of significant digits is often found in applications such as stores parts, where clerical reference is made to the code and where consequently the code can be clearly shown to tie up with the article that is retrieved. One advantage is that it can be changed and extended easily.

Significant digit codes are sometimes referred to as faceted codes, because the component groups making up the overall code number may be thought of as describing a different facet of the individual item. When a faceted code has a free form, in that each facet can take as many characters as required, the resulting numbers may be complicated for data processing. This disadvantage can be overcome by fixing the limits of each facet to a prescribed number of digits. This method has an added systems advantage. For example consider the following code structure, which may be found in a raw material store:

(1) *Facet 1*   Cross-section (single digit)
   1 = Round
   2 = Square
   3 = Hexagonal, etc.
(2) *Facet 2*   Material (two digits)
   01 = Brass
   02 = Mild steel
   03 = Stainless steel
   04 = Wood, etc.
(3) *Facet 3*   Length in centimetres (always three digits; if less than 10, prefix with two zeros; if 10–99, prefix with one zero)
(4) *Facet 4*   Type of finish (two digits)
   00 = In raw state
   01 = Cast
   02 = Machined
   03 = Planed, etc.

Under such a code, a plain metal pin might be numbered 10204502. It would then be uniquely identified and the code could be used as a descriptor in a data processing system. The advantage which arises from a code of this type which classifies each item by its attributes comes from the scope it offers for variety reduction. For example, this item now known as 10204502 might have previously been described as Part 123 (Pin), Part 394 (Dowel), Part 871 (Valve Roller) in this same store. Such occurrences are very common, and duplicate stocks of an item are often carried through inefficient coding.

A garment manufacturer's code might have as its facets, type, size, style, and cloth. The code for this application might, for a given garment, look like

SU M 38L 17 384   Suit, male, size 38, long, style 17, material 384
SO G 08S 02 017   Socks, girl, size 8, short, style 2, material 017

In this example note the mnemonic aids. These make a clumsy code but could be a great help in a busy, cluttered shop. Faceted codes can have each facet in the form of any of the codes mentioned so far — sequence, block, or significant digit.

Another coding system, decimal or hierarchical coding, uses similar principles to the group classification codes, but introduces a decimal point to assist in identifying the major concepts. Consequently, such codes are capable of unlimited expansion by the addition of lower subsets. When a decimal code is used in a data processing application, however, the maximum number of digits must be predetermined. If this is not done, the design of files and subsequent retrieval from them becomes a very complex programming task.

The most obvious example of a hierarchical code is the Universal Decimal Code (UDC) which is widely used for the classification of books. This divides all knowledge into more and more detailed categories, as it descends in the following way:

Code	Items
3	Social Sciences
37	Education
372	Elementary
372.2	Kindergarten
372.21	Methods
372.215	Songs and Games
372.215.6	Action Songs

The UDC code can be extended by the use of linkage symbols, such as a hyphen, an equals sign, and so on, to indicate degrees of relationship between separate code numbers. For example, heat treatment would be coded 621.785 and steel 669.14. We would then know that a document coded 669.14–621.785 is concerned with the heat treatment of steel, because the hyphen linking the two parts signifies that the document deals with the first subject from the point of view of the second.

Decimal codes are sometimes used in commercial applications. An example might be found in the coding of the component parts of a mechanical assembly. As the article is expanded into its subassemblies and then into its component parts, each lower level has decimal digits to identify it.

Alphabetic derived codes are of limited use in commercial applications but are sometimes found in information retrieval applications for document indexing, and large-scale name and address work. They are formed from the original alphabetic version of the title or name, abbreviated by the application of some standard set of rules. Examples of such rules might be to remove all vowels from surnames, or to use the initial letter of the name followed by three digits formed from the second, third and fourth consonants in it. For the latter, the consonants are divided into six groups, thus

Code No.	Letters included
1	B, F, P, V
2	C, G, J, K, Q, S, X, Z
3	D, T
4	L
5	M, N
6	R

The letters W and H are ignored, and Y is treated as a vowel. This list may appear arbitrary, but is in fact based on phonetic principles. Certain rules have to be followed in the application of this system, and it has met with fairly wide acceptance amongst users concerned with this problem of alphabetic coding. An advantage of these types of codes is that the reduction in the number of letters recorded reduces errors due to misspelling. The disadvantages are a certain clumsiness and lack of uniqueness. With the numeric system, for example, the surnames Johnson and Jenson would both be allocated code number J525.

Although not directly concerned with coding in the classification sense, the address generation technique used to index random access files is relevant to the overall subject. The reader is referred to Section 5.7 where the technique is described.

4.10 Check digits

A check digit is, usually, a single digit that is attached to a numeric data item in order to make it self-checking. The check digit is constructed in such a way as to have a unique relationship with the rest of the number.

The construction of the check digit is all important to the type of error that will be detected by it. Take this simple example:

> We have the code number 315, and we form a check digit by adding the individual digits of the number together, thus $3 + 1 + 5 = 9$. The check-digited code number becomes 3159. And

Modulus	Range of weights that may be used	Max. length of number without repeating weight	Weights used	Percentage errors detected				
				Transcription	Single transposition	Double transposition	Other transposition	Random
10	1-9	8	1-2-1-2-1 1-3-1-3-1 7-6-5-4-3-2 9-8-7-4-3-2 1-3-7-1-3-7	100 100 87.0 94.4 100	97.8 88.9 100 100 88.9	Nil Nil 88.9 88.9 88.9	48.9 44.5 88.9 74.1 44.4	90.0 90.0 90.0 90.0 90.0
11	1-10	9	10-9-8 . . . . 2 1-2-4-8-16 etc.	100 100	100 100	100 100	100 100	90.9 90.9
13	1-12	11	Any	100	100	100	100	92.3
17	1-16	15	Any	100	100	100	100	94.1
19	1-18	17	Any	100	100	100	100	94.7
23	1-22	21	Any	100	100	100	100	95.6
27	1-26	25	Any	100	100	100	100	96.3
31	1-30	29	Any	100	100	100	100	96.8
37	1-36	35	Any	100	100	100	100	97.3

Fig. 4.10.1 Efficiency of check digit methods

when writing or punching this number if we transcribe any digit incorrectly, say 4159, the number will now fail our test: $4 + 1 + 5 = 10$ *not* 9.

This type of error is very common in data processing. But if we were to write the number as 1359, transposing the first two digits, another very common error, the number still checks: $1 + 3 + 5 = 9$. Thus when designing a check-digit system the type of errors that are to be detected must be considered. Remember that in data processing, information is frequently punched into cards or paper tape from handwritten documents. Common types of error are as follows:

(1) *Transcription*, where the wrong number is written completely, e.g. writing '1' for '7'.

(2) *Transposition*, where the correct numbers are written but their positions are reversed, e.g. '2134' for '1234'.

(3) *Double transposition*, where there is an interchange of numbers between columns, e.g. '21963' for '26913'.

(4) *Random*, which is a combination of two or more of the above, or any other error not specifically listed here.

Obviously, simply adding the numbers together will not provide a very good error detection method. Most check-digit systems employ weights and a modulus, which are defined as follows:

(1) *Weight*, is the multiplier used on each digit in the original code number to arrive at a product.

EFFICIENCY OF MODULUS 11 SYSTEM-ANALYSIS OF 1000 ERRORS

Number of entries—100 000. Total number of errors—1000 (1%)				
Type of error	No. of errors	Per cent efficiency	Errors detected	Errors undetected
Transcription	860	100	860	0
Transposition	80	100	80	0
Double transposition	10	100	10	0
Random	50	91	45	5
			995	5

Efficiency—five errors undetected in 100 000 entries, i.e. 99.995 per cent coverage of erros

Fig. 4.10.2 Survey results for a check digit test

(2) *Modulus*, is the number which is used to divide the sum of the weighted products to arrive at a remainder.

These terms are explained most easily by a simple example. Consider forming a check digit for a five-figure code number, using the weights 6–5–4–3–2 and modulus 11. If the original code number is 31602, the method of allocating the check digit is as follows:

(1) Multiply each digit in turn by its corresponding weight. This gives

(3×6)	(1×5)	(6×4)	(0×3)	(2×2)
18	5	24	0	4

(2) Sum the resultant products, i.e. $18 + 5 + 24 + 0 + 4 = 51$.

(3) Divide the sum by the modulus and note the remainder, i.e. 51 divided by 11 = 4 and remainder of 7.

(4) Subtract the remainder from the modulus, and the result is the check digit. Thus the final step is $11 - 7 = 4$, and the new code number complete with the check digit becomes 316024.

To confirm that the new code number is correct, it should be multiplied by the chosen weights, using 1 as the weight for the check digit itself; add the results and divide by the modulus. There should be no remainder. The modulus 11 system with weights from 10 down to 2 is probably the most common one in data processing systems and its efficiency varies with the occurrence of random errors. It will detect all other types of error. It is not the only type of system available, however, and Figure 4.10.1 compares the efficiency of various ways of forming check digits. In an investigation within an installation employing the modulus 11 method, 100 000 code entries were examined. The results of this survey, to show how many errors the check digit method found, are given in Figure 4.10.2.

5 File Design

By this stage the relevant output and input designs will have been completed, resulting in specifications using the installation's standard documentation. The next step is to decide how the data is to be structured and physically stored on backing storage devices. Any set of accessible records in a computer system may be described as a file. This chapter looks at types of files, the way they may be organized and accessed—and the impact of database concepts.

5.1 Types of file

To do a job of work a clerk in an office needs access to much more information than he can carry in his head. For this purpose written records of information are needed; they may take several forms. A ready reckoner may be needed by a pricing clerk whose job it is to extend invoice details. Customers' names, addresses, details of recent orders, special discount arrangements, etc., may be needed by an order clerk. The computer suffers from the same problem as the clerk—the amount of information it can hold in its internal memory is relatively limited. So to do the jobs it is programmed to perform it needs access to much more information—which is held on bulk or backing store media like discs and magnetic tape. Each ordered set of accessible records is given a unique name in the computer system.

The basic logical structure of a computer file is as follows, starting with the smallest useful piece of information—an alpha-numeric character:

Character
|
Data item, or field (one or more characters)
|
Record (one or more data items)
|
Subfile (one or more records)
|
File (one or more subfiles)

It is useful to know that files can be categorized to describe the kind of computer processing function they perform. The commonest types are as follows:

(1) *Master file* In a business application these are important files because they contain the essential records for maintenance of the organization's business. They remain in existence throughout the life of the system and are the focus of much of the processing activity. Not surprisingly, it pays to ensure that these files are well designed.

A master file can be further categorized. It may be described as a *reference* master file, in which the records are static or unlikely to change frequently (e.g. a product file containing descriptions and codes; a customer file containing name, address and account number). As the name suggests the file is referred to for information but only occasionally does it need to be amended. Alternatively it may be described as a *dynamic* master file, in which the records are being frequently changed (updated) as a result of transactions or other events. It is of course possible for both reference and dynamic data to be kept on the same file—within each individual record—as with a sales ledger file containing such reference data as name, address, account number as well as dynamic data such as current transaction details and current balance.

(2) *Transaction file* This may also be called an *input* file. Such a file contains data relating to the business activities (e.g. sales, purchases). Its main purpose is for updating a master file.

(3) *Transfer file* This type of file carries data from one processing stage to another (e.g. a transaction file may be input to a sorting process; the output file of sorted records constitutes a transfer file for input into the next process—probably to update a master file).

(4) *Work file* This is a transient file used temporarily for working purposes (e.g. during the course of a magnetic tape merging process to put records into sequence, a number of work files may be used to produce the final transfer file).

(5) *Output file* This contains information—usually extracted from one or more master files—for output from the system. It may be in printed form for despatch to customers or on a magnetic medium for input to another computer process (e.g. invoices may be printed and sent to customers; weekly sales summaries may be output into magnetic tape for input into a monthly accounting process).

(6) *Dump file* This is a copy of computer-held data at a particu-

lar point in time. This may be a complete copy of a master file to be retained to help recovery in the event of a possible future corruption of the master file—or it may be part of a program in which a possible fault is being investigated.

(7) *Archival file*   This is for long term storage of information about the organization's business.

(8) *Library file*   Usually refers to files containing application programs, utility programs and system software generally.

5.2 Storage devices

In theory there is a wide selection of devices and media available for file storage—from punched cards to high speed internal memory. In practice the analyst is concerned only with bulk storage devices using magnetic media such as magnetic tape and exchangeable magnetic discs. Magnetic drums and fixed discs are generally used for special purpose files (e.g. software, tables, indexes); magnetic card devices, although offering huge storage capacity, are slow and not so reliable as other bulk storage devices; punched cards and paper tape are rarely considered suitable for bulk storage of master files.

Magnetic tape units commonly use reels holding plastic tape 0.5 in wide and 2400 ft in length; there is a magnetizable coating on one side of the tape. For reading or writing the tape is transported from one spool to another—like an audio tape recorder. It is impractical to read from and immediately write on to one reel of tape, therefore to update a file it is necessary to read the existing file from one reel and to create an entirely new version by writing on to another reel. When dealing with a master file updated from time to time, the file being read is said to be the brought-forward file and the newly created one is the carried-forward file (because the former has been brought forward from a previous updating process and the latter will be carried forward to a future one). Furthermore, these different versions of the master file are referred to as 'generations' and may be retained for recovery purposes in the event of later versions being corrupted. Three generations (grandfather, father, son) are commonly created and retained before re-using the oldest version.

The magnetizable surface of the tape is densely packed with spots of magnetism measured in bits per inch (bpi) along the length of the tape. They are arranged in rows across the tape—commonly either seven or nine spots per row—each spot representing a bit. These arrangements are also referred to as seven- and nine-track tape. Each row (or frame as it is also called) represents a character or byte. Data is recorded in blocks for which the manufacturer specifies a maximum size, say 4K bytes. Between each block is an interblock gap (IBG) of about 0.6 in to allow the tape to be stopped and started. Depending on the relative sizes of blocks and records one can have one record per block, many records per block, several blocks per record. Depending on the relative sizes of files and reels one can have a single-reel file, a multi-reel file, a multi-file reel.

The capacity of a reel of tape depends upon the length of the tape, the packing density of the data rows and the block size. For example:

Tape length = 2400 ft
Block size = 1600 bytes
Packing density = 800 bpi (bits per inch, which is equivalent to bytes per inch)
IBG = 0.6 in

Then

$$\text{Length of one block} = \frac{1600}{800} = 2 \text{ in.}$$

Add the interblock gap, making a total of 2.6 in per block. Then the number of such blocks on the reel is given by

$$\frac{2400 \times 12}{2.6} = 11\ 076.$$

Thus the number of bytes of data on the reel is $11\ 076 \times 1600 = 17.7M$ approx.

The transfer rate of data to and from the central processor and a magnetic tape unit depends upon the tape transport speed, the block size, the packing density, the IBG, and whether the tape is stopped and started between each block. For example:

Tape transport speed = 200 in per second
Block size = 1600 bytes
Packing density = 800 bpi
IBG = 0.6 in

Then

Theoretical transfer rate = $200 \times 800 = 160K$ bytes per second

Assume tape does not stop, then effective rate is calculated as follows:

$$\text{Length of one block} = \frac{1600}{800} = 2 \text{ in.}$$

But for every 2 in of tape carrying data there is an IBG of 0.6 in, therefore the *effective* transfer rate is given by

$$\frac{200 \times 800 \times 2.0}{2.6} = 123\text{K bytes per second (approx).}$$

Clearly, if the block size is increased, more effective use is being made of the tape and the transfer rate will be higher; similarly, the capacity will be greater. The effective transfer rate is also dependent upon channel utilization and speed.

Exchangeable magnetic discs are mounted in packs containing a number of discs (e.g. six) about 20 in in diameter; 10 of the surfaces are magnetizable. Each recordable surface is served by an electro-magnetic read/write head which can be positioned to record several hundred tracks of magnetizable spots representing bits of data—typically there are 200 such tracks. Conceptually such a disc pack comprises 200 cylinders (numbered 0–199) each with 10 recording tracks (numbered 0–9). This concept is important when trying to minimize access time for records held on a disc drive, because all 10 tracks can be read sequentially without having to move the record-ing heads (which all move in unison). This has a bearing on file organization. A track is subdivided into blocks or buckets—which are the units of transfer to and from the central processor.

A block may be a complete track or part of it—but choosing the block size is very important because if the block size does not divide exactly into the track size there will be wasted space. Discs have interblock gaps, like magnetic tape, but only for block identification purposes, not for stopping and starting. Unlike magnetic tape, discs do not stop—they spin constantly when in use. Depending on the relative sizes of blocks and records one can have one record per block, many records per block, several blocks per record. Depend-ing on the relative sizes of files and packs one can have a single-pack file, a multi-pack file, a multi-file pack.

The capacity of a disc pack depends upon the number of record-ing surfaces, number of tracks, packing density and the block size. The transfer rate of data to and from the central processor and a disc drive depends upon the rotational speed, packing density and blocking arrangement. In both cases simple calculations can be made similar to those for magnetic tape.

The time taken to access data on a disc is made up of the following: (1) the time to position the read/write head over the appropriate track (arm movement time); (2) the time to rotate until the required block is under the head (rotational delay); (3) the time to read or write the required block. For example, a disc drive has a rotational speed of 2400 rpm, therefore the average rotational delay in milliseconds (ms) is given by

$$\tfrac{1}{2} \times \frac{60\ 000}{2400} = 12.5 \text{ ms.}$$

The time taken to read a block of one-quarter track size is

$$\tfrac{1}{4} \times \frac{60\ 000}{2400} = 6.25 \text{ ms.}$$

The time to position the heads varies with their starting position and the number of tracks to be traversed, but an average may be sufficient for approximate timings, e.g. 7.5 ms.

Some manufacturers offer one fixed and one removable pack (also known as a cartridge or module) sharing the same spindle; the former holds system software, the latter application files, etc.

5.3 Serial organization of files

This is the simplest organization of records within a file. The file is created by placing one record after another as it becomes appropriate to file away a record. No regard is paid to keys (identifiers), and if magnetic discs are being used it is possible to achieve almost 100 per cent utilization of the storage space because additional records are merely added to the file—spaces do not have to be left to cater for insertions.

Serial organization may be implemented on magnetic tape or magnetic discs; it may be suitable for (1) transaction files; (2) print files; (3) dump files; (4) archival files; (5) temporary working files; (6) any relatively small file.

There is no logical relationship between the keys of adjacent records and no way of knowing the whereabouts of a particular record, therefore file access to select a record is by a serial search. Average access time is one-half of the time it would take to search the whole file.

5.4 Sequential organization of files

This is probably the most natural and common way of ordering records within a file—by sequencing records according to a key item or a number of key items. For example, the records in an inventory file may be sequenced in ascending order of stock reference numbers.

Sequential organization may be implemented on magnetic tape or magnetic discs. It is extensively used for master files in batch processing systems where there is no demand for immediate access to records. Transaction, insertion and deletion records are batched and sorted into the same sequence as the master file to minimize processing time. In a typical run of a sequentially organized file on magnetic tape, records become available at something like 10 ms intervals, on average; at 15 ms if held on disc. By comparison, if the file were searched randomly it would take say 2 min on average on tape and say 100 ms on disc to find a record. It therefore follows that sequential organization leads to fast processing of master files —provided there is activity on a high proportion of the records. Clearly, if the activity is very low the need to process every record on every run would be time wasting—and there are more suitable methods of file organization (*see* later sections).

Figure 5.4.1 depicts a typical computer process for updating a sequentially organized master file. Transaction records are validated by a data vet program to ensure that only good data is subsequently processed. For example, a transaction record purporting to be depatch information might be rejected in an invoicing data vet run because no quantities had been shown against the items sent to the customer. Such rejections would normally be printed out on an input error report, quoting details of the invalid data.

Valid transaction records are output by the data vet program to form a transaction file.

This file is then sorted to match the record sequence on the brought-forward file. Transactions are matched one-for-one against the brought-forward file, and paired records are read into working storage. The transaction records are used to update relevant fields of the brought-forward record, and the updated version is output to the carried-forward file. At this stage more errors could occur: for example, transaction data might be input for a non-existent master file record. Such errors would be reported on a File Amendment Report. If no transaction record is present, unmatched records are copied from the brought-forward to the carry-forward file un-

104 Basic Systems Analysis

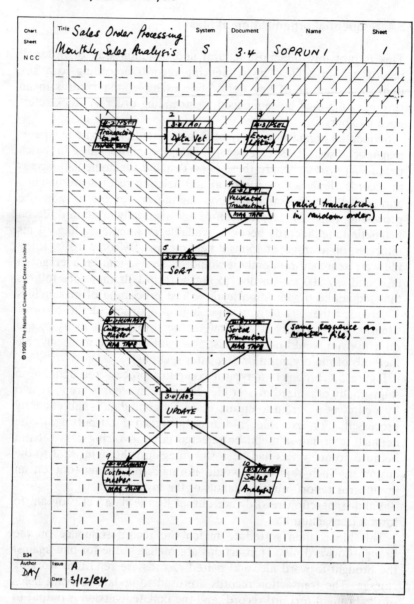

Fig. 5.4.1 Typical sequential file updating process

changed. The carried-forward file becomes the brought-forward file in the next processing run.

Note, in this example, the need to sort the incoming data after validation into the same sequence as the records on the main file. If this were not done, the whole main file would have to be read to process each individual transaction, and then rewound to the beginning for the next transaction, and so on. Such a procedure would be extremely wasteful in processing time.

Of the two common bulk storage media, magnetic tape is usually preferred to discs if sequential organization will satisfy all the user's needs. The reasons include the security provided by magnetic tape aganist file corruption (by facilitating recovery through the use of an earlier generation of the file). It is usually too expensive to use discs like magnetic tapes. However, if the file activity is very low the use of discs will permit 'skip-sequential' (also known as 'selective-sequential') processing to avoid reading inactive blocks—but this requires the use of pointers to locate the positions of blocks of records. If many, small, sequential files are employed, the use of discs may save magnetic tape handling time.

When dealing with sequentially organized files it is a good design aim to try to maximize the activity rate. One way of doing this may be to update the master files less frequently or to partition the master file so that instead of updating the whole file every day only one-fifth is updated each working day—the partitions relating to the natural daily transaction activity. In many applications a small proportion of records account for a large proportion of the activity (e.g. 20 per cent of records accounting for 80 per cent of activity). In these circumstances the records to be changed frequently can be split off into a separate file and updated more frequently.

5.5 Direct-access file organization

There are many applications for which the user demands immediate access to records, for retrieving information about the record or for updating the record. Such operational requirements are associated with on-line systems, real-time systems, immediate or rapid response systems. Direct access to a stored record means that the user can get to the record within a few seconds without having to institute a search through a file, inspecting and rejecting records until a key match is found, as is the case with serially and sequentially organized files. Not all on-line systems offer direct-access facilities to stored master-file records—it may be that data is input on-line during normal working hours but updating of master files is done in batch processing mode overnight.

There are two commonly used methods of organizing files to enable records to be directly accessed: (1) indexed sequential; (2) random.

5.6 Indexed sequential file organization

Magnetic discs are the appropriate storage medium for this method of organization; magnetic tape cannot be used.

Before loading a file on to discs it must be sorted into ascending order of record keys. Also a decision must be made about the volatility of the file—especially its likely expansion, since expansion occurs as a result of record expansion or new records being inserted in the file. There are three stages in making provision for expansion:

(1) Allow space on each track.
(2) Allow space on each cylinder (cylinder overflow).
(3) Set aside extra cylinders (independent overflow).

Such spaces are referred to as overflow areas, and as expansion occurs the track overflow will be used first, then cylinder overflow and finally the extra cylinders.

As the file is loaded on to the discs indexes are created for use in file accessing. Indexes may exist at a variety of levels but would normally comprise (1) disc-drive index; (2) cylinder index; (3) track, bucket or page index.

The disc-drive index merely indicates which drive holds the appropriate pack; it is a small index for which room can easily be found in main memory. The first cylinder on a pack typically carries the cylinder index to point to the appropriate cylinder; the first track on this cylinder carries the track index. Each one of these indexes is a limit index—the highest key is stored for each track in the track index and for each cylinder in the cylinder index.

To access a record, the indexes are inspected one by one to point to a particular track or bucket, etc.; then the track, etc., is copied into main memory where the contents can be scanned to identify the particular record required. The records within a track, etc., may be of variable length and may not be strictly in sequence. If they are in sequence and a track contains a large number of records, a binary-chopping search technique may be used. For example, to find record 23:

Records and keys	13	14	17	20	22	23	24
Inspection sequence				1st		2nd	

The principle is to inspect the mid-point record in successive 'chops' — moving up or down the sequence depending on whether the key to be found is higher or lower than the one last found. In the above example only two inspections are needed instead of six for a sequential search.

Unfortunately, volatile files tend to grow and as soon as cylinder and independent overflow areas are used retrieval becomes slower. To restore operational effectiveness a method must be derived to direct the computer to the overflow record. One such method is known as chaining; this is implemented by inserting the address of the overflow track in the 'home' track. When a record cannot be accommodated in the home track it is placed in the first available overflow track and this track's address is recorded in the home track's chaining address location. When the overflow track itself overflows the same procedure is adopted.

It will be obvious that with a volatile file some degradation of performance will result from the use of overflow areas as more seeks and reads have to take place. When this occurs it is time to re-organize the file. This entails dumping the file on to tape or discs, re-ordering all the live records, reloading and recreating the indexes.

Indexed sequential file organization is versatile. It supports a direct-access requirement yet permits rapid sequential processing. Depending on the fineness of the index, selective or skip-sequential processing is possible, to a greater or lesser degree.

Indexed sequential file organization is well established and well supported by system software.

5.7 Random file organization

Magnetic discs are the appropriate storage medium for this method — like indexed sequential.

This method results in records being stored in locations derived from the record keys. A record is subsequently accessed by derivation of its address by performing the same calculation, etc., on the key. The main advantages of random file organization are that (1) no indexes are required; (2) it permits fastest access times; (3) no sorting of master files or transaction records is required; (4) it handles volatile files well.

The main problem is to devise an algorithm that will achieve a uniform spread of records over the available physical storage medium. Perfect uniformity may not be possible in practice; this results in some areas of store being devoid of records while others will contain more than one record—these latter are referred to as synonyms. If a track or bucket becomes full then an overflow technique must be used to accommodate the excess.

A typical method of generating a random address is by prime number division of the record key—using the largest prime number below the number of available tracks; the remainder of the division is the relative track address. Alphabetic keys or elements of keys can be converted to a number by simply assigning values 1–26 to A–Z. So the key 7H321 becomes 78321. This method of organization and accessing is suitable for real-time transaction processing requirements.

5.8 Other methods of file organization

Other methods hinge upon different methods of indexing. Indexed sequential file organization is an example of the use of *partial indexing*, which involves different levels of index but does not provide an index entry for every record.

With a *full index*, however, there is an index entry for every record giving the location address. The index is strictly in sequential order of keys, although the records themselves may be stored randomly. The main problem with full indexing is the size of the index. However, if size is not a problem a variation of this method is to hold one or two high activity data fields along with the key in the index. These fields can then be accessed at high speed without having to acess the full record on backing storage.

Self indexing is another method, but of limited appeal. In this method the record key is the address. The drawback with it is that it may well give rise to many unacceptable or unrequired key codes.

Yet another kind of file organization is that known as the *inverted* file. In an uninverted file one finds a succession of records each with a number of data items. If one wanted to know the number of occurrences of a particular data item with a specified value, then every record on the file would have to be read and examined. With an inverted file, for each data item one has a list of keys of those records possessing that item; a count of occurrences can quickly be made. There are many variations on this theme — full inversion, partial inversion, secondary indexing and bit map indexing — all of which require considerable study by an experienced analyst to determine their effectiveness and suitability.

5.9 Sorting and merging

Several times in this chapter references have been made to the sequencing of records. In fact the ordering of data items and records is common to almost all aspects of data processing. Ordering of large volumes of data is usually done by sorting and merging techniques. First, strings of data or records are sorted into order and then the ordered strings are merged into longer strings of ordered data or records. In batch processing systems the amount of sorting and merging (usually simply referred to as 'sorting') can account for around 40 per cent of total computer processing. It is important, therefore, that these operations are done efficiently.

In strictness, sorting and merging techniques are the province of mathematicians; similarly, the design and production of programs to sort and merge is a specialist programmer's task — not something for the business analyst to concern himself with.

5.10 Factors affecting file design

It is very important to ensure that the best file design is chosen. Master files are expensive to create and maintain and they are central to the satisfaction of user requirements when the system is up and running. So errors in file design may cost a lost of money to put right, as well as causing damage to the organization's business and reputation through failure to cope with the end-user's needs.

As always there are many considerations and constraints; the ideal may not be attainable and it becomes a matter of maximizing the efficiency and effectiveness of what is practicable in the circumstances. There are a number of things to be borne in mind:

(1) *Operational purpose* Clearly, rapid-response systems dictate a direct-access method of file processing, which in turn means the use of discs and not tapes as the storage medium. At the other

extreme, for daily, weekly and monthly processing needs, where file activity is high, but there are no *ad hoc* or immediate demands for information, then simple sequential organization on tapes would no doubt be the most economical solution. In between these extremes only careful modelling of possible alternative solutions will reveal which is to be preferred — by modelling is meant careful calculations of file capacities, run timings, hardware and software requirements and cost. As an example, a straightforward batch processing requirement with a *low* activity rate may well be more economically processed on disc with indexed sequential file organization than on tape with sequential organization, simply because with the latter the entire file always has to be processed even for one or two transactions. If there are five tapes holding the file, it will take 15–20 min of computer time even though there may only be one amendment per reel! Whereas five amendments on an indexed sequentially organized disc file may take only a few seconds.

It must also be borne in mind that discs are versatile — one can do everything on discs that one can do with tapes *plus* direct-access processing if required.

(2) *Hardware* The analyst ought to ascertain whether there are any constraints on design very early in the investigation. For example, if an existing installation is tape-based and there is an embargo on capital expenditure of any kind, then there is little point in proceeding with a design requiring a direct-access facility — although the analyst may be able to cost-justify his proposals on the *rental* of disc devices in order to circumvent the embargo.

(3) *File size* Small files are probably best kept on disc to improve availability. If many small files are contained on one reel of tape the ones near the end take several minutes to access. A file can be held more economically on tapes than discs — the raw cost of tape storage may only be one-tenth of that for discs — so, other aspects apart, a very large file ought to be on tape not discs.

(4) *Output requirements* An output file held on disc permits greater versatility of processing to meet *ad hoc*, selective requirements — for printing, for example.

(5) *Input requirements* A transaction file is normally subjected to validation, control and sorting. Only a detailed analysis of all the processing requirements in these areas will reveal whether tape or discs will best meet all the requirements.

(6) *File density* Serial and sequential file organizations may approach 100 per cent of the practical capacity of either tapes or discs. With indexed sequential organization overflow must be catered

for unless the file and its individual records are unlikely to expand.

So in practice a balance has to be struck between packing density, access time and file re-organization. Self addressing theoretically achieves 100 per cent packing density—but only if use can be made of all storage locations as keys. The packing density with randomly organized files depends on the distribution of the records as a result of using the chosen algorithm for address generation. The packing density of an indexed sequential file can be specified at file loading stage; this cannot be done with a randomly organized file—although a similar effect could be produced by allocating more initial storage and a higher prime number division if using this method.

(7) *Block density*   On magnetic tape, maximizing block size minimizes stopping and starting time and increases the reel capacity. On disc there is an interblock gap mainly for boundary identification; unless the processing of a block is extremely minimal the head will have traversed the interblock gap long before it is told to read the next block—there will then be rotational delay before the next block can be read. So maximizing block size minimizes rotational delays and increases disc capacity. On the other hand, for low hit rates on disc-based sequential files, large blocks may mean that it takes longer to transfer, unpack and select and read a required record than to process the record; so perhaps smaller blocks would result in faster processing—or maybe indexed sequential file organization would be more efficient.

As a trainee analyst will by now have begun to appreciate, optimizing file design for capacity and processing is no mean task and is more the province of an experienced computer system designer.

5.11 File definition

NCC standards for computer file and computer record specification include the use of the two forms shown in Figures 5.11.1 and 5.11.2. The first of these, the Computer File Specification, specifies the contents, organization, size and use of any file to be read or produced by the computer, but specifically excluding computer printout and machine-readable documents. Each distinct file requires a separate specification, but a single file held at various times either in sorted or unsorted state requires only one specification with both states shown. For each computer file specification there

Computer File Specification	File description		System	Document		Name		Sheet
	Outstanding Orders		MSOP	4.4		OUTORD		1

NCC

File type		File organisation
Input ☐	Master ☑	
Output ☐	Transfer ☐	

Storage medium			Single ☐	Retention period	Number of generations	Number of copies
Mag. tape ☑	Disc ☐	☐	Multiple ☑	3 DAYS	3	1

Recovery procedure

    See operations manual SOP CF1

Keys

    PRODUCT CODE (OSORDT) within ORDER NO. (OSORDH) within CUSTOMER REF

Labels

    Standard

Level	Record name/ref.	Size	Unit	Format	Occurrence
GA	OSORD				3560 - 11323
B	OSORDH	24	W	F	1 (PER ORDER)
B	OSORDT	5	W	F	Av 9. (PER ITEM)

Block/batch size		Unit of storage		Number of blocks	
Actual, for fixed length	Maximum, for variable length	Records ☑		Average	Maximum
	Unblocked	Words ☐		–	–
File size		Bytes ☐		Growth rate	
Average	Maximum	Characters ☐		10 % per ANNUM	
80,928	105,113	Cards ☐		or determining factor	

MAG. TAPE ONLY	Tracks 7 ☐ 9 ☐	Recording density	Speed	Length MAX

DIRECT ACCESS ONLY	Addressing accessing Method	Packing density %	Frequency/condition of re-organisation
	Level	Type of overflow	Size of overflow areas

Notes

    Orders remain on file until despatched or altered

S42

Author	Issue	A
HV	Date	10/12/84

Fig. 5.11.1 Computer file specification sheet

Record Specification NCC					
Record description	System	Document	Name		Sheet
Outstanding Order Item	MSOP	4.7	OSORDT		1

Medium: MAG TAPE
File specification refs.: OUTORD
Record format: Fixed ☑ Variable ☐

Ref.	From	To	Level	Name (In system design)	Data Type	Size	Picture	Value Range
1	0	2	02	PRODCODE	C	3	A99	1-1200 range
2	3	28	02	PRODDESC	C	25	X(25)	
3	29	33	02	UNITPRCE	C	5	Z(*)	1-9999
4	34	37	02	QTYORD	C	4	99	01-99
5	38	39		DISCNT	C	2	X	0-99
6	40	40		PRCODE	C	1		
7	41	44		PARTDEL	C	4		

S 44
Author: DAY
Issue: A
Date: 20/11/84

© 1969. The National Computing Centre Limited

Fig. 5.11.2 Record specification

114   Basic Systems Analysis

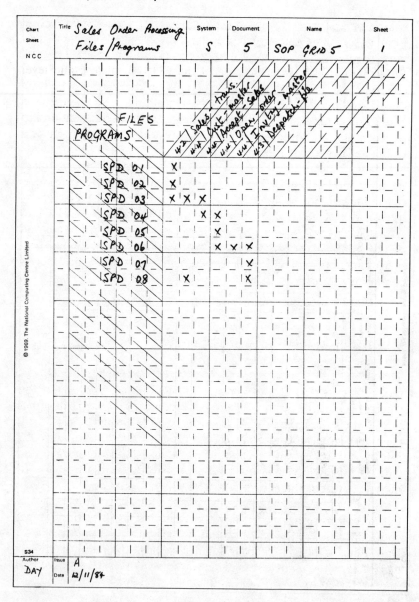

Fig. 5.11.3 Completed grid chart for referencing files and programs

will be one or more record specifications. The record specification form is the same one which was used to specify both input and output records. The notation used in completing this record specification form makes extensive use of the COBOL picture and level concepts. We have already seen how NCC grid charts may be used during the investigation and analysis stages, and these same charts may also be used at the file design stage to show the relationship between files and programs and files and records. An example of a completed grid chart is shown in Figure 5.11.3.

Finally, the specification may include some reference to main memory and backing storage requirements, although the analyst has only a limited interest in these. He is concerned only in so far as any look-up tables which form an integral part of his system may have to be held in memory, and he must leave room for the program and working data. These problems will normally be resolved by analyst-programmer liaison and their agreed solution noted in the specification for reference.

5.12 Database management systems

The terms 'data' and 'information' are often used interchangeably as though they referred to the same thing. This is incorrect and in Chapter 1 a distinction was made between data and useful information. Data is language, mathematical or other symbols which are generally agreed to represent people, objects, events and concepts. Information is the result of modelling, manipulating and organising data so that it increases the level of knowledge and the ability to take decisions of the people who have the information.

At its lowest level data is represented by bits, characters or bytes, but in systems terms the lowest level which we need to consider is a data item. This is the smallest unit of data that can have meaning. Data items may have a fixed length—typically numeric data—or have a variable length as in the case of an address. A data item is often referred to as a *field*. Groups of data items that have a close association with each other are referred to as data aggregates; day, month and year aggregate to date. Records are groups of data items and perhaps aggregates that relate to a single entity. Records are identified by a key field or fields: a unique key identifies a record specifically, a generic key identifies a type of record. This is shown below:

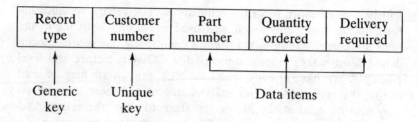

Record type	Customer number	Part number	Quantity ordered	Delivery required

Generic key   Unique key   Data items

Individual data items do not exist in isolation; there are relationships between them. Relationships can be one to one as in the relationship MARRIAGE between husband and wife, one to many as in the relationship TEACHES between lecturer and students, and many to many as with STOCKS between warehouses and parts. Examining a data item such as CAR we can provide its make, model, colour, etc. The data item is known as an *entity* and make, model, etc., describe the *entity* and are known as *attributes*.

Traditionally, applications have been developed in isolation with the consequent development of separate files of data specific to each application but with the same field occurring in several files: part numbers could occur in sales order processing, stock control and purchasing systems and in field service and maintenance systems. The problems of keeping the value of such data items consistent are considerable. This situation worsens as each new application is developed.

The alternative database approach however sees the organisation as a co-ordinated and systematic whole where data items are stored only once and never duplicated. Database management systems — as opposed to single file application based organisations — are the most appropriate data handling technique in these cases. Databases and database management systems offer five advantages:

(1) Redundancy of stored data is reduced. Different applications needing the same data item can be given access to it and the item is stored only once.

(2) Inconsistency of data is eliminated. With data stored only once, inconsistency cannot arise.

(3) Stored data can be shared. New applications can use existing data.

(4) Data can be made more secure. Database management systems provide access control for individual users as well as to individual data items.

(5) Data independence is provided. The separation of the data

from the way it is processed means that changes in an applications programme do not affect other programs that access the same data. Equally, changes in the way the data is physically stored do not affect programs.

Essential to the successful use of database management systems are a database administrator and a data dictionary. The database administrator (DBA) is responsible for all operational data. The main responsibilities are:

(1) Deciding the information content of the database
(2) Deciding the storage structure and access strategy
(3) Defining the authorisation checks and validation procedures
(4) Liaising with users about data
(5) Defining the strategy for backup and recovery
(6) Monitoring performance

The data dictionary contains descriptions of all entities and their relationships. It will also hold information about which programs use which pieces of data. The data dictionary is crucial to the DBA in maintaining the overall system.

6 Systems Design

From his consideration of input, output and files, the analyst will discover the processing and computational requirements that are necessary to provide the links between them, and a new system design will be born. It is in this area that the skill and experience of the systems analyst play a vital role. He alone has the responsibility for developing the design of the new system, of explaining his design to his management, and securing their approval.

It is in this second step—of providing an explanation—that all the various systems designs within an installation will possess a common denominator. While the analyst is exploring alternative solutions to the problem at hand he is very largely a free agent, and his own ingenuity controls his activity. Once the solution has been resolved and it is necessary to communicate it to programmers and users his freedom is restricted. The restriction is imposed by the documentation methods employed within his installation. Even though the standards for documentation may vary widely from one installation to another, it is essential that only one method be employed within any one organization. Without this standardization, communications will become chaotic and overall efficiency will suffer.

This chapter first deals with the traditional approach to system design. It examines common problems of design and then discusses the establishment of a standard method of documenting the agreed design. The chapter ends with a brief look at the more recent concept of structured programming and, by implication, structured design.

6.1 Considerations in system design

When the analyst gets down to the task of designing the actual processing and computation of the data in his system, he will face numerous problems. Many of these will be due to constraints imposed by the user or to limitations in the hardware and software that make up his installation. In the notes which follow, a selection of the matters requiring attention is listed. It is not possible to be comprehensive, as the variety of likely problems is so great, and no two installations are exactly similar. However, the comments made will give the inexperienced analyst a feeling for some situations he may encounter during the design stage.

(1) *One-input systems* may be a small routine or a large suite of programs. The stage at which a given function becomes large enough to justify making it a separate procedure is usually determined by the amount of coding involved. Such systems may arise, however, when the function is particularly complex and/or requires specialist knowledge *or* where the appropriate routine or program already exists in usable form.

(2) *Two-input systems* almost always imply a matching procedure. This may be by a serial method when both input streams have been put in the same sequence, or by direct access. In the latter case, records will be available on a direct access medium with short access time. Transactions are dealt with on arrival and not necessarily in a predetermined sequence. This method is obligatory in a real-time system. There is a variation of the serial method, which is to take several 'bites' at one file and for each such 'bite', pass the whole of the other file. Each 'bite' is held in store (or, possibly, on a direct access medium) and each pass of the transaction file processes all transactions applying to the records held in the current 'bite'. It may be better to rewrite the transaction file during each pass, dropping off those transactions which have been matched. This method can clearly save on sorting time. Note that, even when it is possible, through the use of direct access media or by the variant described above, to deal with one or both files in random sequence, this may be an erroneous or unacceptable procedure because there is a definite order for several transactions applying to the same record (e.g. one must amend a record before using it to answer an interrogation). In this case, the transaction file must be sorted.

(3) *Multiple-input.* Generally, it is not convenient to handle more than three inputs in a single program. In particular cases, however, it may be that a multi-input matching procedure is an attractive method. Often the number of peripheral devices on the computer will establish the limit to the number of input streams which can be handled. Where the method is to be 'two inputs' then a system of merges or multi-input sorts will have to be used. Careful consider-

ation may need to be given when deciding whether to mount a large sort-merge system immediately before the two-input run or to sort and merge the several input files progressively as they become available, e.g. daily data capture and validation process leading to a weekly main run.

(4) *Spooling and media conversion programs.* 'Spool' = simultaneous peripheral operations off-line (tape-to-printer, cards-to-tape, etc.). The use of these programs will normally impose certain limitations on format and/or sequence. These limitations are usually well worth accepting in view of the benefit obtained from the use of proven programs and the advantages of standardized operating procedures when handling the slow peripherals.

(5) *Multi-programming.* The operating system may impose quite significant constraints on the way in which multi-programming facilities are used. Thus a 'foreground-background' system with a partitioned store may be the basis on which the machine has to work or a full 15-program system may be available or a multi-access system allowing, say, 200 remote users to work concurrently. Even within the manufacturer's constraints, the installation operating management may impose restrictions of their own, designed to ensure that the facilities of the machine are profitably exploited.

(6) *Real-time working.* Here input, output, and process with access to files must be handled immediately and each procedure is performed once only before going on to another: work is by transaction rather than by process (as in batch processing). The approach is usually to deal with all types of input transaction in one message-handling program. This is an input program which accepts, vets and examines the message. The message will define the procedures required and the files to be accessed. The message-handling program then deposits the decoded message data in a standard place and initiates a transfer of control to a processing program. It is this program, or sequence of programs, which deals with the message and then deposits the results in another standard place. An output program then picks it up and passes it to the remote terminal. The processing program may have passed on details of the message for further processing. It is usual for the manufacturer to provide programs, or sets of routines from which programs can be assembled, to handle the communications tasks. These programs will impose certain standards of internal communications on the customer programs. The communications package will deal with context-independent checking, line failures, queueing problems, etc. — all of which are specialist topics.

(7) *Run segmentation*. The segmentation of the total task is the crucial stage in system design. The basic task is to identify the division of labour between computer and non-computer. Generally, the smaller the number of programs the better. This is, however, a gross over-simplification made essentially from an operational point of view. It will usually be necessary to segment because the program would otherwise be too large to be accommodated in store, and/or the necessity for a sorting function implies a split in procedures (e.g. Data Vet-Sort-Update).

(8) *Optimization*. In Section 1.3 it was stressed that the design process is iterative. Having completed the initial design scheme, the analyst must examine the proposed design for weaknesses and pay particular attention to questions of peripheral or processor-domination and file design. The most straightforward methods of optimization are those of run combination and run segmentation. Combination of runs reduces set-up time and may enable one or more passes of a file to be saved. If a computer-limited run can be combined with a tape-limited run, this may result in a balanced combined run. Segmentation of runs will tend to increase set-up time but can be used to reduce demand on peripherals. In particular cases of processor-bound runs, it may be possible to effect a reduction in processing time by partitioning into two balanced runs or by moving a particular processing function into the next program.

(9) *File design* must be examined carefully to ensure that the right degree of compactness has been achieved. If one or more runs handling a given file are file-limited, it may be worth attempting to reduce the physical sizes of the file records, so reducing file passing time. Such compacting will tend as a secondary effect to increase the processing time, since extra coding will have to be inserted in the programs concerned to unpack the compacted fields. Conversely, where runs are processor-bound, it may be worth expanding the records, so easing the task of unpacking the record fields, in an attempt to obtain a balance. Where a system is to be run on a multi-programming machine, of course, it is highly desirable to avoid processor-bound runs, since this will tend to limit the flexibility of the system from the operational point of view. Activity distribution on files should have been considered from the beginning and any '80–20' situation (where 20 per cent of the main file records receive 80 per cent of the transactions) should have been detected and the file design modified accordingly (*see also* Chapter 5).

(10) *File buffering* may or may not be subject to control by the programmer. Where it is possible to nominate one, two, three or

more buffer areas for a file, this may be of use, particularly where a list requires significant processing time but the activity of the file is very low. A large number of buffers can then be employed with advantage to maximize the amount of overlapped file-time, by allowing a queue of blocks to build up while the list is being processed.

(11) *Sorting runs* will almost invariably use the standard sort program supplied by the manufacturer. The use of the first and last pass 'own-coding' options should always be borne in mind. In an extreme case, this can save a whole run. If the mechanics of the manufacturer's sort program are fully understood, it may be possible to arrange that the data to be sorted is produced by the preceding program in such a way that the first pass of the sort is made easier and faster. Similarly, if the first pass of the sort is processor-bound, a reduction in the storage used by this pass in forming strings may bring the pass into balance. In critical situations, a pass in Phase II of the sort might be saved by reducing record size and hence *increasing* string length out of the first pass.

Many of the contraints and problems mentioned above are programming-oriented. This is deliberate. They have been included to impress upon the analyst how necessary it is to work closely with the programming team during the design stage. If the analyst ignores the programming implications of his design he does so at his peril, because he is likely to find that many of his pet ideas and theories are not feasible on the equipment at his disposal. A senior programmer, if consulted early enough, when the theory is in embryo, will quickly point out those ideas which will cause programming difficulties. By such liaison, the analyst will save himself much wasted effort.

6.2 Design flowcharts

After the analyst has recorded all the relevant facts, he will study them and, as a result of his study, will produce a tentative design for the new computer-based system. This first design will be discussed and probably modified as a result. After several drafts have been discussed, the analyst will be in a position where he can specify his system in a suitable manner for programming work to begin. At all the stages in this process the use of standard documentation will be of great help.

The most common form of representation of data processing functions is the flowchart. The systems analyst uses flowcharts to plot the flow of information in and around his system. The procedure flowchart has been mentioned (Chapter 2) and, depending on the standards for his installation, the analyst will also draw design flowcharts each at a different level. He uses standard symbols, and represents the various aspects of the system with a particular symbol.

The same structure of documentation which was used to record the existing system during the fact-finding stage in Chapter 2 can be used at the systems design stage. There are four levels of systems design flowcharting. The first of these is the system outline, which shows inputs, outputs, processes and files independent of sequence of operations in a way which is quickly and clearly understandable both to DP staff and to non-DP staff. The system flowchart next depicts in outline the sequence of events in a system showing the department or function responsible for each event. The name of each department or function is entered on the top of the grid, inputs are shown on the left, outputs on the right. Processes are shown by the appropriate symbol in the relevant column. The symbols are connected by lines and arrows, each symbol containing plain language annotation. The computer run chart depicts the interrelationship and, where relevant, sequence of the computer routines to be performed, showing inputs, files and outputs. This form of chart is always required in the design of the computer aspects of a new system. Finally, the computer procedure flowchart depicts the sequence of operations and decisions in a computer procedure.

Figures 6.2.1–6.2.4 inclusive show completed examples of these flowcharts. The flowchart symbols and their meanings used in the NCC Systems Documentation Standards are shown in Chapter 2.

The main principles to be observed in drawing flowcharts are as follows:

(1) Make the flowchart clear, neat and easy to follow. It will then make a good visual impact and communicate well. This implies (a) marking the logical start and end points; (b) using standard flowcharting symbols; (c) avoiding crossed flow lines; (d) using simple decisions, i.e. those giving yes/no or greater/equal/less than answers; (e) working in a consistent direction down or across the page.

(N.B. Draw a rough plan before starting the final chart so as to get the boxes arranged advantageously.)

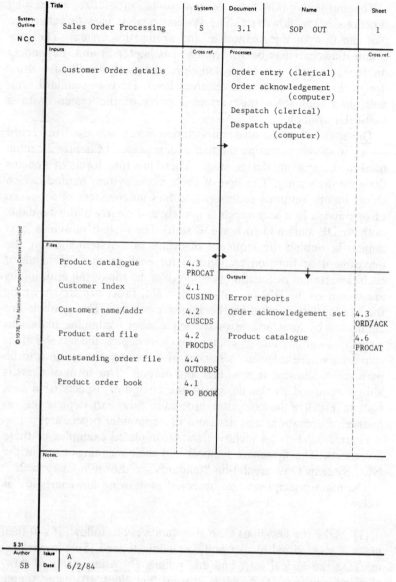

System Outline N C C	Title Sales Order Processing	System S	Document 3.1	Name SOP OUT	Sheet 1
Inputs		**Cross ref.**	**Processes**		**Cross ref.**
	Customer Order details		Order entry (clerical) Order acknowledgement (computer) Despatch (clerical) Despatch update (computer)		
Files					
	Product catalogue	4.3 PROCAT	**Outputs**		
	Customer Index	4.1 CUSIND	Error reports		
	Customer name/addr	4.2 CUSCDS	Order acknowledgement set		4.3 ORD/ACK
	Product card file	4.2 PROCDS	Product catalogue		4.6 PROCAT
	Outstanding order file	4.4 OUTORDS			
	Product order book	4.1 PO BOOK			
Notes.					

S 31

Author SB	Issue Date	A 6/2/84

© 1976. The National Computing Centre Limited

Fig. 6.2.1 System outline

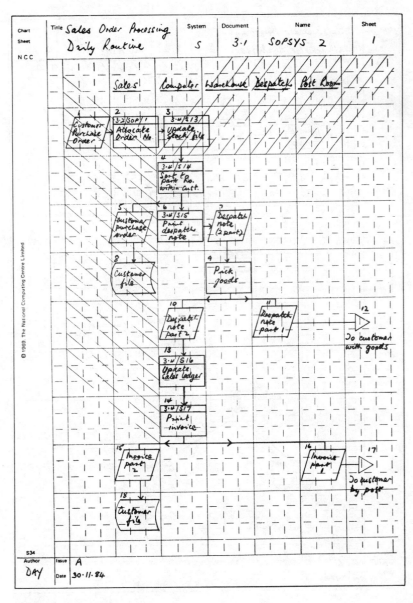

Fig. 6.2.2   System flowchart

126 Basic Systems Analysis

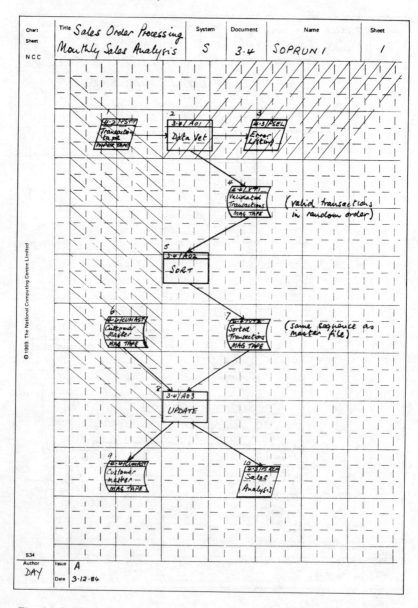

Fig. 6.2.3 Computer run chart

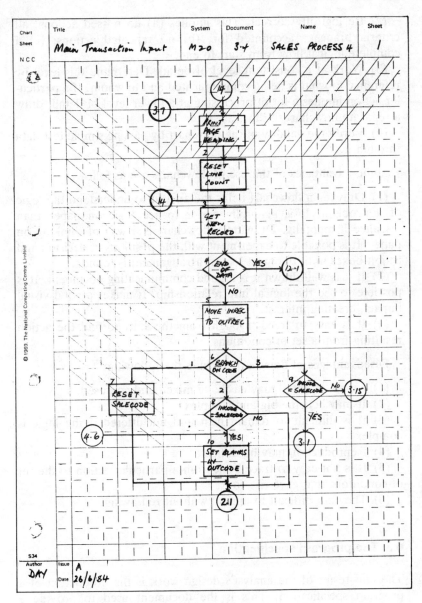

The flowchart shown contains the following:

Header fields: Chart Sheet NCC; Title: Main Transaction Input; System: M20; Document: 3.4; Name: SALES PROCESS 4; Sheet: 1

© 1969. The National Computing Centre Limited

14

3.7

1 PRINT PAGE HEADING

2 RESET LINE COUNT

14

3 GET NEW RECORD

4 END OF DATA — YES → 12.1

NO

5 MOVE INREC TO OUTREC

6 BRANCH ON CODE — 1, 2, 3

7 RESET SALECODE

3 → 9 INKODE = SALECODE — NO → 3.15
YES → 3.1

4.6

8 INKODE = SALECODE — NO
YES

10 SET BLANKS IN OUTCODE

2.1

S34 | Author: DAY | Issue: A | Date: 26/6/84

Fig. 6.2.4 Computer procedure flowchart

(2) Be logically correct. This implies (a) no missed actions, all criteria branches accounted for; (b) no repeated actions; (c) a complete logical flow from start to end.

(3) Work at a consistent level of detail. Determine this level by reference to what is being communicated and to whom. If a particular action implies a lot of processing at a lower level of detail, draw a separate flowchart to illustrate it.

(4) Verify the validity of the flowchart by passing simple test data through it.

Some advantages of the use of flowcharts are as follows:

(1) Drawing a flowchart helps the analyst to understand the logic of a problem. It is generally easier to see logic on paper than visualize it mentally. Drawing a flowchart is a way of thinking on paper. It is possible to experiment with the positions of decision and action boxes until the most logical arrangement is found.

(2) If it is well drawn it is a good communicating document. It is possible to see the logical interrelationship. It makes a good visual impact, creating interest.

(3) It is easy to trace through from the start to find the action resulting from a set of conditions.

Disadvantages of flowcharts include the following:

(1) Levels of detail may easily be mixed and confused.

(2) It may become big and complex.

(3) It is difficult to remain neat and uncluttered if the logic is complex.

(4) Reproduction may be difficult.

(5) It is not easy to trace back from a particular action to the set of conditions that gave rise to it.

(6) They are difficult to amend without redrawing.

6.3 The program specification

The end result of the analyst's design work is the production of a program specification. This is the document used to provide a communications link between analyst and programmer. It must be meticulous in its detail, because this communication must be perfect. The programmer will look to his specification for guidance on the innumerable questions of logic that will arise as he draws up his program. No loose ends will be allowed.

The actual programming of the system may take many man-months. By the time a program trial reveals some inconsistency in the basic design, the analyst is likely to be immersed in the details of a completely different system. Unless his specification is clear and correctly documented, his chances of discovering the fault and suggesting a remedy will be remote. Another reason why program specifications must be unambiguous is because of the nature of the analyst's job. His work will often take him away from the computer centre and into user areas of the company. If queries arise about his specification while he is absent, it is unlikely he will be able to resolve them by telephone. Therefore, valuable programming time can be lost while the problem awaits the analyst's attention.

The methods of presentation employed in the program specification are likely to vary from one installation to another. They will depend to a large extent upon the personalities of the analysts involved and their depth of experience. Whatever the actual presentation method is, it is vital that it be standardized. The forms presented previously are suggestions for standards, and these will be supported by a narrative. The check list which follows indicates the details which need to be included within a good program specification.

SECTION A — PRELIMINARY INFORMATION

1. *Title and responsibility*
   (i) System name
   (ii) System number
   (iii) Date produced
   (iv) Produced by (name, address, telephone)
   (v) Produced for (name, address, telephone)
   (vi) Origination of amendments
2. *Authorization*
   (i) Statement of acceptance
   (ii) Signature of person accepting
   (iii) Area of responsibility
3. *Contents list*
4. *Amendments record*
   (i) Amendment number
   (ii) Authority reference
   (iii) Date of amendment
   (iv) Section number

(v) Page number
(vi) Description
5. *Time schedules*
    (i) First test with live data
    (ii) Parallel or pilot running commences
    (iii) Solo running commences
    (iv) Other systems to be linked
6. *Definition of terms*

SECTION B — AIMS OF THE SYSTEM

1. *Procedures covered*
2. *Departments concerned*
3. *Adjacent procedures*
4. *Benefits*
    (i) Financial
    (ii) Manpower
    (iii) Control
    (iv) Information availability
    (v) Information quality improvement
5. *Outputs*
    (i) Medium
    (ii) Description
    (iii) Use to be made
    (iv) References to specification section
    (v) Volumes
6. *Inputs*
    (i) Medium
    (ii) Description
    (iii) Information source
    (iv) Reference to specification section
    (v) Volumes
7. *Files to be maintained*
    (i) Medium
    (ii) Description
    (iii) Use to be made
    (iv) Reference to specification section
    (v) Volume

SECTION C — DETAILED SYSTEM DESCRIPTION

1. *Clerical procedures*
    (i) Flowcharts

(ii) Narrative, including procedure descriptions (daily/weekly/monthly activities); codes; controls; data transport and handling

2. *Data preparation output distribution procedures*
   (i) Flowcharts
   (ii) Narrative, including procedure descriptions; codes; controls; data transport and handling

3. *Computer procedures*
   (i) Flowcharts
   (ii) Narrative, including no. of runs (daily/weekly/monthly); purpose of runs; function of programs.

SECTION D — CHANGEOVER PROCEDURE

1. *File conversion*
   (i) How data is collected
   (ii) Editing required
   (iii) Codes required
   (iv) Data preparation
   (v) Programs required
   (vi) Runs required

2. *Pilot running*
   (i) Data to be used
   (ii) Volumes to be handled
   (iii) Processing period
   (iv) Expected results

3. *Parallel running*
   (i) Processing period
   (ii) Anticipated volumes
   (iii) Check points

4. *Direct changeovers*
   (i) Changeover period
   (ii) Overtime to be worked
   (iii) Extra staff to be employed
   (iv) Anticipated delays

SECTION E — EQUIPMENT UTILIZATION

1. *Equipment specification*
   (i) Type of computer
   (ii) Peripherals
   (iii) Ancillary equipment

2. *Equipment utilization (computer)*
    (i) Number of runs
    (ii) Frequency of runs
    (iii) Set-up time
    (iv) Estimated time per run (minimum, maximum, and average volumes)
    (v) End of run procedure times
    (vi) Total computer time for daily, weekly, monthly, etc., procedures

3. *Equipment utilization (ancillary equipment)*
    (i) Machine type
    (ii) Type of activity
    (iii) Frequency of activity
    (iv) Time for each activity (minimum, maximum and average volumes)
    (v) Total equipment utilization time for daily, weekly, monthly, etc., procedures

SECTION F — SOURCE DOCUMENT SPECIFICATIONS

For each document
1. *Description*
    (i) Identification, i.e. name, number, purpose
    (ii) Method of origination, i.e. how and where filled in
    (iii) Elements of data, including field names, description (alphanumeric/numeric); field lengths; maximum values; origin of each element
    (iv) Frequency
2. *Document sample*

SECTION G — PRINTOUT SPECIFICATIONS

For each printout
1. *Format planning chart*
    (i) Printout reference number and name
    (ii) System name and number
    (iii) Program name and number
    (iv) Number of print lines per sheet
    (v) Maximum field sizes
    (vi) Lateral spacing requirements
    (vii) Vertical spacing requirements

2. *Narrative*
   (i) Purpose
   (ii) Type of stationery (preprinted, blank, size, quality)
   (iii) Special features required (e.g. use of limited character set)
   (iv) Distribution and routing
3. *Samples*

SECTION H — FILE SPECIFICATIONS

1. *For all files*
   (i) Medium
   (ii) name
   (iii) Labels
   (iv) Size (maximum/average)
   (v) Number of record types
   (vi) Sequence
   (vii) Block size (unit of transfer)
   (viii) Fields per record, showing field names; field descriptions; field lengths; maximum values; permitted signs
2. *For card files*
   (i) Card types
   (ii) Number of card columns per field
   (iii) Punching code
3. *For paper tape files*
   (i) Interblock gaps
   (ii) Block start and end signals and field markers
   (iii) Punching code
4. *For magnetic tape files*
   (i) Tape sentinels
   (ii) Packing density and gap size
   (iii) Number of reels
   (iv) Frequency of use
   (v) Retention period
5. *Direct access files*
   (i) Storage method (random, serial or sequential)
   (ii) Storage access method (including system, address computation)
   (iii) Bucket size
   (iv) Number of buckets

SECTION I — DESCRIPTION OF SYSTEMS TEST DATA

1. *Input data*
   (i) Listing or layouts
2. *Main files*
   (i) Layouts
3. *Expected results*
   (i) Logic results
   (ii) Arithmetic answers
   (iii) Line printer results (Format Planning Chart)

SECTION J — PROGRAM DESCRIPTIONS

For each program
1. *Introduction*
   (i) Program name
   (ii) Program number
   (iii) Purpose
2. *Start procedures*
   (i) Standard
   (ii) Non-standard
3. *Main procedures*
   (i) Processing requirements
   (ii) Validity checks
   (iii) Control totals
   (iv) Maximum, minimum sizes
   (v) Signs
   (vi) Rounding requirements
   (vii) Error conditions
4. *End procedures*
   (i) Output of control totals, tables and analyses
   (ii) Dumping requirements
   (iii) File closing
   (iv) Operator messages
   (v) Entry to other programs
5. *Dump and restart procedures*
   (i) Halt points
   (ii) Dump medium
   (iii) What is to be dumped

When the program specification is handed over, the chief programmer or a senior programmer will divide the work up for the

PROGRAM CONTROL SHEET

Page ————————

Date ————————

Program [   ]   Analyst ——————   Supervisor ——————————

Computer system flowchart: Input/output links

Activity	Programmer assigned	Start date	Finish date	Actual days
Review specification Flowcharting Coding Program testing Operating instructions				

Fig. 6.3.1 Program control sheet

actual writing of the programs to take place. A Program Control Sheet, as shown in Figure 6.3.1, will be made out for each program so that progress can be monitored. Although not strictly a systems analysis document, brief details of this document are given here for completeness. The control sheet states for each program an unambiguous program name, and any links with other programs within the application. Target dates are quoted for completion of the program logic flowcharts at various levels, of program writing, and testing, and for completion of operating instructions.

When the program specification has been completed, the analyst has partially finished his task of introducing a complete system. A number of other jobs remain, however, and these form the subject of Chapters 7–11.

6.4 Design strategy

The system design process begins with a functional specification of what the new system is to achieve and ends with a detailed system specification from which programs can be prepared. A well-designed system has the following characteristics:

(1) *It achieves the goals set for it.* In day-to-day operation, systems produce a wide variety of outputs for different users. The frequency of these reports ranges from 'daily' to 'annual'. It is therefore very easy to confuse the production of these reports with the goals of the system. At the beginning of the system design process it is very important that we do *not* fall into this trap. A 'system goal' is something that the system will achieve, that is unlikely to change, and that can be expressed in business terms. We shall come back to this shortly when we consider how to define system goals.

(2) *It reflects the structure of the business application it serves.* Business application systems usually follow well-defined, structured processes. This is particularly the case in accounting systems for example, where legal constraints and traditional business practice formalize the way in which accounting processes work. In the lifetime of a computer-based system, however, it would be unreasonable to suppose that business demands would remain static and that no changes would be made to the designed system. A well-designed system is therefore able to cope with business change because it is capable of being modified easily. This can often be done when the

system design reflects the business procedure and where computer processes mirror the real world.

(3) *It is secure, accurate and reliable.* Controls and security are covered in more detail in Chapter 7, but careful design ensures security against the incorrect use of data, privacy from unauthorized access and a complete audit trail and provides the ability to recover completely after system malfunctions.

(4) *It is easy to understand.* Simple solutions are to be preferred in every case. A complex solution which optimizes the use of some technical aspect of the design is more difficult to specify, program, test and modify.

The design process is an iterative activity. As each of the design elements is considered, a re-examination has to be made of the design decisions taken so far and an assessment made of the relationship between these previous design decisions and the element being considered. This repetitive activity continues until every aspect of the proposed system has been considered and the final and complete design has been reached. The basic steps in the overall design process are:

(1) the definition of the system's goals
(2) the preparation of a conceptual model or logical system design
(3) the development of a physical system design
(4) the preparation of the system specification

In defining the system's goal it is important to remember that this is not the same thing as providing specific users with specific outputs. We can illustrate the difference between goals and user requirements by taking an accounts receivable system as an example. The goals of an accounts receivable—or sales ledger—system are:

(1) To maintain an accurate and timely account of debts owed to the organization by its customers. To provide internal audit controls that will ensure that debt records are accurate and that the system performs reliably.

(2) To provide appropriate management information to support the organization's overall objectives, such as the control of debt and the establishment of customer credit-worthiness.

By definition, these goals are not likely to change. It is the content and format of the inputs and outputs which are variable and which good system design will allow to change. On the other hand, a system designed to produce a specific output format will need to be redesigned each time the content of that format changes.

Preparing a conceptual model or a logical system design requires that the analyst determines how the system will achieve its goals. This takes account of the logical inputs and outputs and of the processes necessary to transform one into the other. No notice is taken at this stage of the organizational constraints within which the system will operate. This is done during physical system design. In the short term, cost performance and reliability figure largely during physical system design. In the long view, the flexibility of the system, the expected growth in usage and its life expectancy may be important: higher performance and reliability may be achieved at a greater cost. In designing the physical system, the analyst must consider the desired performance from the system, its cost and reliability; development time, maintainability and flexibility; potential for growth and life expectancy.

Out of these conflicting requirements will be derived the optimum system capabilities which enable the logical system goals to be achieved within the organizational constraints within which the system must operate.

7 Controls and Security

A computer system is made up of four parts:

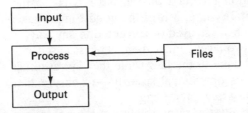

Each and every part of the system can, and almost certainly will, give rise to errors. Designers must take this into account. They must design their system so that it can detect when things go wrong.

Any mistakes must be spotted as early as possible. The system must prevent small errors being compounded into serious problems.

There may also be a requirement to provide controls for auditors. The designer must keep in touch with the internal audit department so that audit requirements can be designed into the system rather than built on at the end.

7.1 Input

Some data in the system is of little importance. However, other data is of fundamental importance while some is open to fraud. We must concentrate on these last two categories.

Input for the system must be received; authorized; input; processed. Most of these points were covered in Chapter 4 but a summary here would not be out of place.

Data control must know what input to expect and when to expect it. They must be able to recognize the signature of the person who can authorize that input. They must make sure that the input they receive is subsequently passed on to and processed by the system.

Input should be submitted in batches with suitable batch header details. These details are recorded by data control. The program which reads the data into the system also prints batch details. These can be checked against the details previously recorded.

Subsequent programs in the system should record records processed to give a continuous tally throughout the system.

7.2 Files

Once a system requires a file, then the problem of file security arises. Since the information recorded on the file is vital to the efficient working of the total system, it is of the utmost importance for this information not to be destroyed or corrupted in any way.

Security is exercised at two levels. First, there is physical security —to ensure no damage is done to the files while they are awaiting processing. Second, there is operational security—to ensure that data are not lost during the actual processing.

Physical security is normally the responsibility of the operations supervisor in an installation. Part of his function is to ensure that suitable housing facilities exist for the files, including fire protection for all master copies. If magnetic media are employed, he will be responsible for keeping them free from dust, in a suitable atmosphere without excess temperature or humidity, and away from stray magnetic fields. He will also make certain that correct files are issued to the operating staff and that proper records are kept on the movement of all files. All tape cans, and other containers, will be suitably labelled to give adequate identification of their contents. The supervisor will also instruct all operators in the correct methods of handling the media to avoid physical damage, such as creased magnetic tapes, or torn punched cards.

All files are given purge, or expiry, dates. A record is kept showing the date on which the magnetic medium can be regarded as new, irrespective of its current contents. This date would normally appear on the file label as an added security check.

Operational security is a joint effort by systems analysts, programmers and the manufacturers of the equipment. The provision of hardware devices by the manufacturers to assist operators process files correctly is one such facility which exists. For example, a 'read' ring can be placed on the tape by the operator which prevents any data being written on to it, as security in a system which requires a tape always to be read. 'Write rings' are used in a similar manner for the reverse situation.

Another common facility is for hardware to perform parity checks on individual characters after they have been processed. The parity checks are a method used to detect invalid bit patterns by the addition of an extra bit to each character. When the sum of the bits which make up the character *plus* the parity bit total to an even number, the system is called even parity checking. A similar method — to sum to an odd number — is known as odd parity checking. The implication of a parity failure is that there has been some hardware malfunction causing the incorrect transfer of data and this will cause the processing to stop and a suitable error signal to be displayed. Before this happens, however, a parity failure causes the program to loop back to an earlier point, recover the original data and attempt the processing again. Devices of this type are given the generic name of hardware protection. These might consist of performing numerical reconciliations after a certain number of records have been processed, and abandoning the run if the checks fail. It is usual to dump the contents of internal storage at the time of the failure, so that the program may be restarted at the same point when the cause of the failure is discovered. In some circumstances, the analyst will specify logical points in the processing where meaningful checks can be carried out. Some care must be taken not to over load the program with checks, however, as restart situations can be expensive even if the files are secure.

When using direct access files, some consideration must be given to what restart action is possible if a processing error is detected. If files are updated by copying (*see* Section 5.4), the problem is not serious as the original data on the old file are still available after updating. With copying by the overlay method of processing, however, more care is needed. As the contents of the old file are overwritten with the new version, they are not available if the new version proves to be incorrect. This problem is overcome by copying the original record into a temporary storage area until the processing has been checked and found correct. If there is an error, the original part of the file is still available and processing can be repeated. If no error exists, the temporary storage area can be used again to copy the next set of records from the old file while they are being updated.

Systems analysts are involved in devising suitable file control totals which can provide useful safety checks. Reconciliation of brought-forward file totals plus transaction totals against carried-forward file totals can highlight errors before the situation is irrecoverable. There are two main types of control total employed for files: (1) *quantity or value totals*—where the totals themselves are meaningful and may be of value in actual processing; (2) *hash or nonsense totals*—that have no meaning for processing, other than for reconciliation purposes. An example of the use of both types might be found in a payments received system in a hire-purchase application. Customers would send their monthly remittances to the computer centre, where the account number and amount would be recorded for input. Batches of input would contain a control slip containing two totals, which would also be punched. One total would be a hash total of the relevant customer account numbers and the other would be a value total of the moneys received. During the updating of the main customer account file, the computer would store the sum of the account numbers of records actioned, the sum of amounts recorded on those records and the current value of all outstanding balances. When the updating run was finished a reconciliation would be possible by (1) comparing the input hash total with the sum of account numbers actioned, to ensure all relevant accounts had been updated; (2) comparing the sum of the amounts recorded with the input value totals to confirm that all moneys had been written to the file correctly; (3) calculating the new total outstanding balance from the old balance on the brought-forward file minus the sum in (2) above, to double-check that the record was accurate.

Another file security technique is known as the grandfather–father–son method. With this scheme, three versions of a file are available at any one time. File 3—son—is the current file which was created from file 2—father—which, in turn, was created from file 1—grandfather—as shown in Figure 7.2.1.

The advantage of this technique is that recovery is always possible. For example, if the new data on file 3 were found to contain some errors, the job could be repeated using file 2 again with the transactions data. If both file 2 and file 3 were damaged during the processing, file 1 is still available to create file 2 afresh, and this can then create a new file 3. If file 3 is found to contain no errors, it is used as the father file in the next processing cycle to create a new son file, which will be physically output on to the old file 1.

Direct-access devices are theoretically more in need of security features and routines since many of the safety measures used with magnetic tapes are absent. For example, several files may be kept on one disc, programs may be kept with other files, updating is often by overlay, and in many cases no grandfather–father–son files are kept. In practice, the direct-access systems may be just as

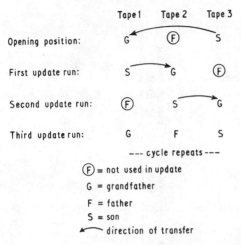

Fig. 7.2.1   Generations of files for security

reliable as magnetic tape systems because hardware checks and software routines are provided by the manufacturer. Also, because the problems are inherent to the medium, both systems analysts and programmers are particularly careful to insert additional checks into all programs employing direct-access facilities.

Apart from the labels used on the outside containers of storage media, there are also file labels incorporated as part of the recorded data itself as a further security check. The label acts as the first record on the particular file. It contains enough information to enable the computer to check that the correct file has been set up. This is done often by some type of date comparison with the current processing date, or by the use of sequential run numbers. Purge dates would appear on the label as well.

On magnetic tape files, the label is usually the first physical record on the tape. With direct-access files, particularly discs, all labels for all files contained tend to be kept together in one place. In addition to the usual checking information required, the label will indicate the whereabouts on the disc of the first working record of the file. The software control features, which are used to prevent unauthorized reading or writing in the same way as the hardware devices on magnetic tapes, would also be found in the label of a direct-access file. Figure 7.2.2 illustrates typical file labels for both tape and disc showing the kind of information normally found on them.

1 HDR 1-4	File name 5-14	File number 15-19	Tape number 20-22	Creation date 23-28	Expiry date 29-34	Generation number 35-37	Systems code 38-50	Spare—for user information 51-80

1 HDR 1-4	File name 5-14	File number 15-19	File start address 20-25	File end address 26-30	Creation date 31-35	Expiry date 36-40	Systems code 41-52	Spare—for user information 53-80

Fig. 7.2.2 Magnetic tape and disc labels

Dumping file contents periodically is another technique used to improve security. This means copying the whole contents on to another physical storage medium which can be housed separately from the working file. This method is applied particularly to direct access files which are being updated by overlay, and to transaction files which may be needed if reconstruction of the main file becomes necessary.

Care must be exercised, however, as dumping can be a time-consuming routine. Magnetic tape may prove to be a good receiving medium for a dumped file, particularly when this can be arranged as an overlap in a routine in which the file is being read from disc. Reading from disc and dumping on to disc may be a more difficult operation, as in some systems it is not possible to process two discs simultaneously, or to read from one and write to another in the same pass.

Duplication of files is a security technique which is not employed a great deal in the United Kingdom, but is one which can be advantageous when working with large files. The method is to make amendments to the main file and to take a copy of the new version on a second device at the same time. For example, by overlaying the main record on disc, and copying the amended recorded on to magnetic tape.

In some cases, complete duplication of files may exist when two exact copies of each file are maintained. This is rare, but could be necessary for extreme accuracy in an interrogation system.

Some physical security measures should also be taken to ensure that the initiating documents are not lost or damaged. If queries arise after the file has been amended, it is necessary to be able to check that the original instructions were followed. Serially numbered documents which are retained on file at the computer installation, or in the originating department, often ensure the required control.

7.3 Process

During the process stage we must rely on hardware and software. Hardware errors will be detected by the hardware itself and are outside the scope of the designer.

Software errors are best avoided by testing programs thoroughly prior to implementation. There are two kinds of trials to be considered: link trials and operational trials.

Link trials are concerned with testing the programs to see that they behave satisfactorily when fed with data prepared by other programs in the suite, as distinct from data prepared by their author.

Pairs linking is the first stage, in which for a given program all programs that feed data into it are identified. The program is then combined with each of these feeding programs in turn to form a series of pairs. Each pair is tested on the machine. The test data used for these trials should be all, or part, of the program test data constructed for the feeding program, the output from which is fed into the program being linked. This output will already exist. Errors in either of the pair of programs may be detected at this stage.

The first technique to be employed for detecting errors during link trials should be to discover what is the essential difference between the data being fed from the feeding program and the original program test data which have already been performed successfully. There are the following four broad categories of difference:

(1) Quantity of data may be different.

(2) File layouts of the same files may differ between the pair of programs.

(3) Although file layouts may be the same, a different understanding of the functions of certain of the fields may exist between the authors of the two programs.

(4) New circumstances may exist in the data which were previously untried.

Having identified the essential difference, it should be possible to recognize the part of the programs affected, and hence the cause.

System linking follows pairs linking. In this stage, the test data prepared by the systems analyst and submitted with the specification are punched and submitted to the suite as a whole on the computer. Before this can be done, it will be necessary to prepare suitable files. These should be prepared using the file amendment facilities of the suite designed for insertions. Again, errors should be detected by recognizing the essential difference between the data now being used and that previously tested.

Even though the suite of programs has been linked and tested with system test data, it would be imprudent to commence operating the job using the computer in place of the old system until the suite is tested for a considerable period under realistic conditions. A large quantity of current, or past, transactions should therefore be fed into the suite.

These operational trials check not only the programs, but also the data preparation and all related manual systems. Errors will be found in the programs, in the system proposed, and, very possibly, in the old system which it replaces. Many staff, including systems analysts, will therefore be involved in this operation. It is very costly as three operations are being performed simultaneously.

(1) The new system is being run.
(2) The old system is being run.
(3) The results of the two are being compared.

Because of this it should be kept short, and all possible should be done during individual program and link trials to ensure that the parts are in good shape before operational trials begin.

Program errors arising should be detected and corrected by the

PROGRAM TRIALS LOG

PROGRAM:

Assembly No.	Trial No.	Link trial No.	Date	Mis-operation	Computer fault	Data preparation	Trial data error	Assemblies		Trials		Link trials		Notes
								Total standard time		Total standard time		Total standard time		
								Elapsed time	Effective time	Elapsed time	Effective time	Elapsed time	Effective time	

Fig. 7.3.1   Program trials log

programmers. The essential difference technique described previously will be helpful in this connection, but, since the programmers concerned will often not be the author, some manual simulation may be necessary.

A Program Trials Log (*see* Figure 7.3.1) should be kept by the programmers for each program. This log records progress made during trials and link trials. Each trial, assembly, or compilation is numbered, using a separate column for each. The date of despatch to the computer is shown. Data preparation, operator, or computer errors are recorded on the log against each trial.

The time recorded on the copy of the operator's log returned with the completed trial is entered under the time claimed. The effective time agreed to be charged for the trial after investigation of the results is entered under time charged. The program name and number are entered at the top of the sheet, and also the standard time for trials, assemblies and compilations.

The results achieved by each trial are recorded on the test data

SUITE: SUITE TRIALS LOG

Date sent	Program	Date performed	Date returned	Time elapsed		Time in query	Adjustments		Remarks
				This trial	Total		This trial	Total	

Fig. 7.3.2   Suite trials log

description sheet by entering the number of the trial that successfully performed the required action on each piece of trial data. These enteries are made immediately following each trial so that progress and the amount of testing still required can always be measured.

In addition, a Suite Trials Log, as shown in Figure 7.3.2, should be maintained by the chief programmer, or a senior programmer in charge of a suite of programs, in order to (1) check the number of trials shown by each programmer in his program trials log is correct; (2) determine the accumulated computer time to be charged for trials at any time; (3) measure the delay between despatching a trial to the computer and its completion and receipt back in the programming office; (4) record the date the trial actually took place; (5) establish an overall daily progress position for all programs.

7.4 Output

Designing output was the subject of Chapter 3. Here we will summarize a number of control aspects.

(1) *Always produce something.* If for a particular run there is nothing to be printed, the user will still be expecting something. Simply print that there are no details for this run.

(2) *Print a termination message.* At the end of a report make it clear to the user that it is the end, e.g. total line or 'end of report' message.

(3) *Print control totals.* These assure the user that everything has been processed and output.

(4) *Print headings and page numbers.* The user needs to know what the report is about. It also helps to print date report produced, and possibly time.

7.5 Audit controls

The owners of the organization are also interested in control and to look after their interests they appoint auditors. The functions of auditors are laid down in certain cases by legislation. They are interested in the detailed controls as it is their duty to examine and report upon the efficiency of the whole system of controls, financial and otherwise. The great difference in the auditors' approach when considering computer systems lies in their inability to examine records which are now on tape or disc, and to check procedures that exist in the form of program instructions.

For this reason, it is necessary in any computer-based system to establish an audit trail, which is the means by which the details underlying summarized accounting data, i.e. control and reconciliation totals, may be obtained; it includes the location of supporting documentary evidence. This trail must be included at the design stage. Management will also be interested in similar data in the event of outside queries, e.g. when invoice and statements disagree with customers' calculations. A trail must go right through the system, from data collection to result distribution, and can often use many of the controls already established. The problem of records, as files change frequently and old versions are destroyed causing a break in the trail, can be overcome by printing out changes and filing the printouts so that a complete external history is available. Another method is to keep additional copies of files for audit purposes, a point which is discussed in Section 7.2 on file security. Procedures exist in the form of a program and unless the auditor is expert he will not understand their technicalities. For this reason, he will be interested in the trails system, and the test data used and results achieved. His interest will extend to the security methods which ensure that the approved version of the program is being used, and the processing log should display program identity, including the version number. The log must also include details of mechanical failures and of action taken.

Some auditors have their own test packs which give known results with a particular set of programs. They keep these test packs in their own office to prevent alteration; when performing an audit, they run the test pack with the program and check that the results agree with the original set. In this way they can ensure that unauthorized program modifications have not been made. Although the theory of this is sound, it is not wholly successful in practice as in day-to-day working many minor program modifications have to be incorporated and these could invalidate the auditor's test data as auditors' visits tend to be at lengthly intervals. Consequently many auditors satisfy themselves that security procedures for the introduction of program modifications are adequate and that authority for these is given at a suitable level

These problems will become more acute with the advent of the real-time systems where no history is kept and control is immediate. Similar complex problems are involved in recording the actions of a self-regulating system with the ability to learn. However, the objectives of the auditor and management are the same and the former will accept a reasonable control system established by management. Thus, the analyst has a considerable responsibility to his own management to provide suitable audit controls within this system.

7.6 Security

Security is the protection of data against accidental or intentional disclosure to unauthorised persons, or against unauthorised modification or destruction of data. Apart from commercial or defence reasons, in the UK the Data Protection Act requires that there is adequate protection for personal data. Security controls and procedures are therefore now a legal requirement as well as an operational necessity. The following essentials should be considered during the design of every system:

(1) Data should be protected from fire, theft and other forms of destruction

(2) Data should be reconstructable. However good security maybe, accidents will happen

(3) Data should be auditable

(4) Systems should be tamperproof. Ingenious programmers should not be able to bypass the controls

(5) Users must be positively identified by the system before they are given access to the data

(6) User's actions should be authorised once they have been allowed to access the system

(7) Actions in sensitive areas should be monitored so that illegalities can subsequently be traced back to individual users

Some simple security countermeasures can be taken. Access control will prevent unauthorised users from gaining access to the system or the data. Methods available now, in particularly sensitive systems, include fingerprint or voice print recognition, although password protection is usually considered to be adequate.

The widespread use of microcomputers poses special security problems, particularly as users of microcomputer systems and standard packages have little or no previous computer experience. Because they are not aware of the need for security hardware, software and data often remain unprotected. There are now known cases of microcomputer theft—hardware, software and data—

where users have paid a ransom to get their property back, not because the hardware was valuable but because the software and data was essential and security backup copies had not been taken.

7.7 Summary

The major factors discussed in this chapter are summarized in Figure 7.7.1 below.

Elements	Danger	Protection	Limitations
File	I/O failure	Copy	Cost of unused copies
		Recreate	Time, source data, sequence
		By-pass	Lost data
			Programming costs
	Wrong version	Check total	Not necessarily unique
		Headers	
		Naming conventions	Exceptions (year end)
		Generation data groups	Re-runs
			Very old versions
	Fire	Off-site storage	Turn round problems
Input data	Missing data	} Batch checks	Balancing errors
	Double reads		Data must be sequenced
	Missing batches	} Batch numbering	Can be restrictive on users
	Double reads		
		File totals	Batches must be sequenced
	Re-run problems	Use a tape copy for input	
	Fire	Use a tape copy for input	
	Invalid data accepted	Program validation	
		Check digits	
		Motivated data originators	
		Check digits	
		Meaningful check totals	
Reports	Not used	Exception reports	
	Missing pages	Print 'end of report'	
	Invalid data	Meaningful check totals	
	Poor quality printing	Eye-checks	

Elements	Danger	Protection	Limitations
Programs (and JCL)	Bugs	Testing Check totals	Never complete Must be correctly positioned in the system or program
	Amendments	Testing	Expensive and ineffective
		Authorization	Never time
		Documentation	Never up-to-date
	I/O failure	Back-up	Often not up-to-date
	Testing	Formal procedures for updating the live library	Can be by-passed Often expensive
Unknown	Unknown	Logs Routine dumps Back-up machines Analysis of SMFs Audits Quality of staff and skills	

Fig. 7.7.1

8 Proving the Design

A systems analyst is also a salesman. Devising the most beautiful system is only part of the job. Someone has to be persuaded to buy it. Someone has to allocate staff, money and time to installing and running the system. The users of the system will need to be convinced that they are going to get the information they need in time for it to be of use. A systems analyst has, therefore, to test and time the systems he devises. Some timing knowledge will also be required to evaluate the equipment manufacturers' claims.

This chapter considers these combined aspects of proving the finished design — testing and timing.

8.1 Testing the system

It is important to test the new system thoroughly before it is implemented, to prove that it will perform the tasks it has been designed to achieve.

The results of testing will prove that the system is working correctly; this will (1) give added confidence to the system designer and his team; (2) inspire the confidence of the users in the new system; (3) prevent holdups and frustration during implementation, and reduce the requirement for maintenance.

Testing is taking place in some form throughout the project. As each step of the new system is designed, the designer should examine it on paper to ensure that, logically, it should perform its designed task.

There are usually limitations to the resources which can be made available for testing. It is necessary, therefore, to plan system testing. Clerical aspects must be tested as well as the computer aspects.

The systems analyst must carry out the following actions:

(1) Produce an outline of the new system (flowcharts, decision

tables, description of inputs, outputs and files, and the specification for each procedure).

(2) Plan the tests (stating who is responsible for producing the item to be tested, who is responsible for testing it, when the test is to take place, to whom the results of the test should be communicated, what action should be taken as a result of the test).

(3) Produce test data (to ensure that the range of values which can be expected are acceptable, that values outside those ranges are rejected and produce the correct error action). Each program specification must define the input data values and ranges which are acceptable and which combinations are not acceptable.

The programmer must

(1) Use the specifications, flowcharts and decision tables supplied by the systems analyst for detailed programming.

(2) Use procedure-oriented and specialized languages which are appropriate for the application system that has been designed.

(3) Desk-check his own decision tables, flowcharts and programs for logical and clerical errors. Then get another programmer to desk-check them.

(4) Send his coded program for punching and verification.

(5) Send his program for compilation, eliminating syntax and compilation errors in succeeding compilations until the source program is free of errors.

(6) Recompile to produce a corrected print of the source program, and the object program on a magnetic tape or magnetic disc.

(7) Run a set of test data against his object program to ensure that it meets its specifications. When fully satisfied the program should be handed over to a quality assurance (testing) programmer.

The quality assurance tester must

(1) Use the test data supplied by the systems analyst to ensure that the specifications for each program have been met, and from the input values will produce the expected results.

(2) Link programs together when program testing on each program is complete and test the linked programs—link trials. This is to ensure that output records of one program will be accepted and actioned by the next sequential program.

When the final computer program has been linked in, the whole system must be tested. The system test must include all areas of data capture, file updating and output, to ensure that known input will produce the expected output.

Areas for testing should include (1) clerical preparation of specimen forms of all types which will be used; (2) transmission of the data; (3) data input control; (4) data preparation and verification; (5) validation of input; (6) error routines; (7) file creation (in preparation for eventual file conversion during implementation); (8) file updating; (9) output control; (10) output handling and distribution; (11) output usage by the recipient (this is to ensure that the recipient correctly understands the output and then takes the appropriate action as a result of it).

Some common points which can be tested by the analyst in this way are as follows. In data vet or validation programs, obvious items to check are the effect of (1) oversize/undersize items; (2) out of range items; (3) incorrect formats; (4) negative numbers; (5) no data at all; (6) invalid combinations; (7) input of corrections; (8) effect on daily and batch controls; (9) batch control testing; (10) error bypass routines.

Input data to updating programs can be designed to examine the effect of (1) input and no file record; (2) file record but no input; (3) wrong record formats; (4) wrong, or out of date, files employed; (5) zero and negative amounts, which may cause calculation trouble.

The final results can be checked to test the effect of (1) wrong stationery; (2) printing halts/restarts; (3) punching errors/restarts.

8.1.1 Clerical procedure testing

Human error is the most common reason for systems to fail. Clerical procedure testing involves manual procedures, such as input document coding, input control and the use of output. Testing should simulate actual operating conditions as far as possible rather than be specially designed to cover every abnormal occurrence. If possible 'live' data should be used in expected quantities. Testing in this instance can also be a training stage since those being tested must first be instructed in the procedures to be followed.

The testing should ensure that (1) the procedures are fully understood; (2) when errors occur correct action is taken; (3) timings are realistic; (4) volumes of work predicted can be handled; (5) the maximum possible peak load will not cause the system to fail.

Personnel used in these tests must be those who will be handling the input documents, error returns, control forms, output, etc., after implementation. It must be realized, however, that testing will

be an extra workload and that, with the old and proposed systems
running at the same time, the effects may distort the results of the
test.

In systems testing, the absence of files not yet constructed often
causes problems and made up files have to be used which may not
contain the true record distribution. This is due to the difficulties of
testing a combination of man/machine procedures which have not
been fully implemented. To help overcome this, use of standard
documents and flowcharts is made to enable the whole system to be
tested. Logic can be tested by this means to ensure that all alternatives have been considered.

8.1.2 Real-time systems testing

Modular programming should be used for real-time systems. Modular programming allows more complex problems to be handled in
small individual programs known as modules. In a batch processing
mode it is common to have a 'mainline program' controlling access
to each module. Modules should be limited to 250 instructions or
one page of core storage (usually 4K). This allows each module to
be programmed and tested independently. For speed and efficiency
in a real-time environment each module should control the flow by
instructing the real-time control program (RTCP) to call the next
module. A RTCP differs from a 'mainline program' in that it does
not control the program flow but contains the housekeeping and is
the first module to be accessed.

The message path through the modules is called its thread. Single-

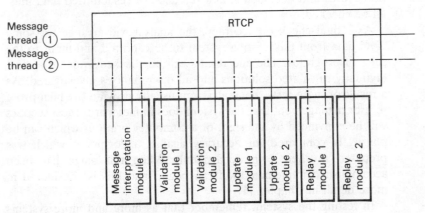

Fig. 8.1.1   Real-time program organization

thread systems will only process one message at a time. Multi-thread systems will process two or more messages concurrently (Figure 8.1.1).

Testing in a real-time environment is greatly simplified if modular programming is used. As each module is coded it is passed to a quality assurance tester. As successive modules are completed they should be linked together and link-trials carried out. Finally, when the last module has been coded, the whole program must be tested. Abnormal input must be tested as well as the normal transactions.

The analyst must recognize the two distinct types of testing with which he will be involved. These are testing the system in abstract, and, in the system's operational aspects, testing which are the concrete facts associated with people and machines. Figure 8.1.2 illustrates the general scheme of introduction of a computer system and shows the two areas of testing involved. In the abstract the test is of logic, information flow, and the design. In the concrete, it is of the efficiency of communication and training. Abstract tests must always precede concrete ones and they can be done in stages.

During the creation phase, program testing in subunits can be performed in parallel with clerical system testing in subunits. Eventually, the time arrives when testing passes from abstract to concrete, thus: (1) testing form design for completion and extraction; (2) testing instructions for completeness and ambiguity; (3) testing the transmission network of data and results, which involves messengers, office clerks, post, railways, etc.

The importance of testing error correction routines based on expected computer output is stressed. Response time, resubmission procedures and documents, and rejection of resubmitted data must all be checked.

As a final check on accuracy, the analyst will arrange for some 'live' data from the existing system to be captured and input to the suite of programs. The end results will be checked in detail against manually calculated solutions and all discrepancies investigated. As a result of these tests the analyst can usually provide the final proof to user management that the system does in fact work. Most sceptics will be convinced by the sight of a computer printout which can be physically compared for accuracy against a document which was prepared in their own department. When this stage has been accomplished implementation can begin, and this is considered in more detail in Chapter 11.

In testing the system, remember that as more and more systems are put on to computers, auditors are becoming more interested in

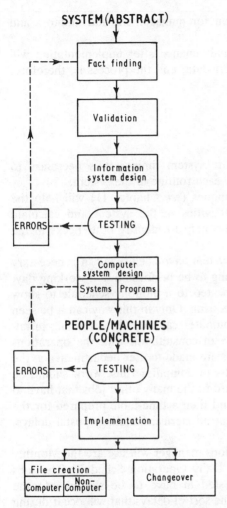

SYSTEM (ABSTRACT)

Fact finding

Validation

Information
system design

ERRORS ←— TESTING

Computer
system design

Systems | Programs

PEOPLE/MACHINES
(CONCRETE)

ERRORS ←— TESTING

Implementation

File creation

Computer | Non-Computer

Changeover

Fig. 8.1.2   Stages in systems testing

this aspect. Take the internal auditor's advice as you may face
having your systems audited by external data processing auditors.

To summarize this discussion of system tests, the analyst must
bear the following points in mind:

(1) Keep in touch with programmers to ensure that their test data
is effective. Efficient testing can save time later.

(2) When the system is handed over, test the complete system.

(3) Maintain documentation for management and auditors and user departments.

(4) Program and system maintenance after implementation will require further testing and test data, and the process is, therefore, continuous.

8.2 System timing

To obtain reasonably accurate system timings it is necessary to calculate the elapsed time for each routine in the system.

Critical path analysis techniques (*see* Chapter 11) will help the analyst to identify all the activities in the system and calculate durations for each. It will also help to identify bottlenecks in the system.

When individual runs in a system have been timed, it is necessary to consider how these are going to be performed in a working day. The total load should be allocated to a monthly schedule to show the effect of peaks in the processing. Only in this way can it be seen whether there is adequate computer capacity. In a running installation this work can be done in consultation with the operations manager. Before any promises are made to user departments on the timeliness of reports or results of computer runs, work should be allocated to days, having regard to the many other jobs that have to be done in the installation, and then a time-table prepared for the processing, taking into account all clerical handling, postal delays, weekends and so on.

The tools which the operations manager will use are the Monthly Operation Schedule and the Daily Operation Schedule. The jobs the systems analyst is interested in have to be slotted into the existing load. Only then can the sort of delays that will occur during production running be seen. Figure 8.2.1 illustrates an operations schedule; the difference between the daily and monthly versions being of time-scale only. These schedules are simply horizontal bar charts and the principles underlying bar chart construction are discussed in more detail in Chapter 11, in the context of timing implementation.

When allocating times to the jobs involved, remember to make due allowance for delays which may occur by the nature of the system: This is particularly relevant when items have to be transmitted to a central preparation unit from outlying stations. Figure 8.2.2 illustrates a typical job for preparing weekly statistical data,

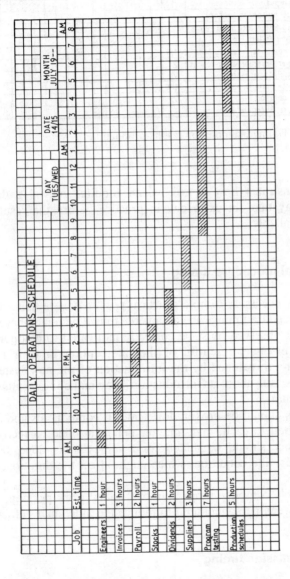

Fig. 8.2.1 Daily operations schedule

Day	Content of job	Hours
Monday	Edit incoming data (80% recd.)	1
	Batch input, prepare controls	1
	Punch input	2
	Verify input	2
	Correct punching errors	1
Tuesday	Edit and batch late input (20%)	1
	Punch/verify late input	2
	Set up computer – vetting runs	1½
	Correct errors from vet: re-submit	2
	Start main processing	½
Wedny.	Complete main processing	1½
	Agree controls	½
	Decollate, burst and despatch	1

Fig. 8.2.2 Timing example for results production

and demonstrates how a task which requires only 1½ days to process may take an extra day in practice before the results can become available.

8.3 Clerical procedure timing

It is usually necessary to hold practical tests to ensure that accurate clerical procedure timings are obtained. It must be remembered that those being observed must be correctly motivated to co-operate in the test. Personnel resentful of the interruption to their normal work, feeling threatened by the introduction of the new system, or over-enthusiastic about its introduction, may make the results hopelessly inaccurate.

Each task should be broken down into its components and a time value assigned to each, man-minutes per form filled correctly, etc. An allowance should be added for complex tasks done seldom, interruptions, and conditions which might affect the duration of the task.

Few people complete more than five hours' work per day. Interactions between personnel tend to reduce the individual's efficiency.

8.4 Transmission timing

Transmission timing can be considered under two headings: physical transmission and telecommunications.

8.4.1 Physical transmission

Physical transmission may be part of a clerical procedure and so may be dealt with as above. More often it is a scheduled collection, either by carrier or Post Office, and delivery times may create a critical bottleneck in the system such that if the schedule is not adhered to the receipt of output may be delayed beyond the acceptable time limit. Contingency instructions should be written into the user manual to take account of a breakdown in the schedule.

8.4.2 Telecommunication timing

For on-line systems, especially in a real-time environment, it may be necessary to calculate the response delay caused by the transmission.

Ideally the number of characters in a message divided by the line speed would give the transmission time. But further delay is caused by (1) transmission errors and the need, therefore, to retransmit erroneous blocks; (2) protocol overhead, the extra characters transmitted for line control; (3) the time taken for a signal to travel the complete return circuit, termed the loop propagation delay (the signal travels at the speed of light and therefore the delay is almost negligible); (4) the time taken for the receiver to interpret the message and assemble an appropriate reply, added to the time taken by the original transmitter to interpret the reply, termed the receiver and transmission delays; (5) the time taken for each modem in the system, if in half-duplex mode, to switch from receive to transmit termed the modem turn-round. This further delay is calculated by the equation

$$\text{Additional delay} = \frac{B}{B + Lc + R(P + D + T + M)}$$

where B is the block length in bits, Lc is the protocol overhead, R is the line speed in bits per second, P is the loop propagation delay, D is the receiver delay, T is the transmitter delay, and M is the modem turn-round delay.

The total time for the transmission can be calculated by the formula

$$\text{Total transmission time} = \frac{\begin{array}{c}(\text{Number of message blocks} + \text{Number of}\\ \text{blocks to be resubmitted}) \times\\ \text{Number of bits per block}\end{array}}{\text{Line speed in bits/s} \times \text{Additional delay}}$$

Further delays may be caused by communication control procedures and queueing.

8.5 Hardware timing

Any discussions of hardware timing have to be generalized. It is necessary to look for a timing expert in your own installation and elicit the facts about the actual machine you have. All that is done here is to explain the principles involved in timing some of the more usual equipment and give a guide as to the questions to ask the expert. In any given installation the timing formulae and factors should be summarized in the Standards Manual for everyone to use as required. The type of central processor affects both instruction times and the number of instructions to be executed. Considerations are summarized as follows:

(1) *Word machines* — Process one word at a time
   — Economic for big numbers
      (36 bits = 36 000 000 000)
   — Arithmetic is very fast
   — Inefficient for small numbers
   — Have to 'unpack' short fields
   — Long fields may bridge word boundaries
   — May have to use double length arithmetic
   — Use more instructions

(2) *Character machines* — Can handle variable length fields
   — Work in decimal
   — Economic for short numbers
   — Ease of data movement
   — Slower

(3) *Byte machines* — Economic storage of decimal numbers
   — Large numbers of special characters
   — Variable length fields
   — Work in decimal or binary
   — Can use registers
   — Can also work from storage to storage

(4) *Instruction sets* — The larger the number of different instructions provided the fewer you will have to solve the problem.

To time the internal processing of a computer run you must know
(1) the average instruction time for an average commercial problem

(example 50 μs); (2) the number of instructions to be executed (only the main loop in the program need be considered; an adjustment can be made for exceptions afterwards if desired). The process time may be obtained by the following formula:

$$\text{Process time} = \text{Number of instructions in main loop} \times \text{Number of transactions} \times \text{Average instruction time in } \mu\text{s}.$$

You should (1) record your guesses; (2) count the instructions in the main loop of some programs; (3) ask experienced people — programmers or systems analysts — 'Is this guess reasonable?'; (4) compare your guess with the finished program.

In timing input and output units, it is necessary to know the operating system, which may be able to read in information or write out information at the same time as processing. The types of operating system available include the following features: (1) no overlap; (2) simultaneous input and output, not overlapped with processing; (3) buffered units which only have to wait for data transfer from a fast buffer; (4) use of several buffers in main storage. However, the job can never be run faster than the slowest unit.

The computer may also have an interrupt system so that the processing can continue regardless of the input and output, but it still takes time. It is estimated that in simple systems it may add a 20 per cent overhead, and in complex systems a 50 per cent overhead. It may be that the central processor is fast and the input/output units comparatively slow, so runs may be input/output-bound in any case.

When considering the input of data to the system from magnetic backing stores, there are other considerations apart from the working speed of the units. The manufacturer's reference manuals will always give full information on this, and Figure 8.5.1 summarizes the formulae for calculating times on tape, disc, drums and magnetic cards.

Peripheral devices are connected to channels and the characteristics of these affect timing:

(1) If you have many tapes on one high speed channel you can only work one tape at one time.

(2) You can have several card readers, punches or printers at once on a special slow speed channel.

(3) Systems using high speed channels may be limited by numbers of channels and cannot read or write two devices at once on one channel.

Media	Formula
Magnetic tape	$\left[\dfrac{\text{Total characters}}{\text{Character rate}}\right]+$ (No. of blocks × Start/stop time)
Magnetic drum	$\left[\dfrac{\text{Total characters in block}}{\text{Character rate}}\right]+$ Rotational delay
Magnetic discs	$\left[\dfrac{\text{Total characters in block}}{\text{Character rate}}\right]+$ Rotational delay + Arm movement
Magnetic cards	$\left[\dfrac{\text{Total characters in block}}{\text{Character rate}}\right]+$ Rotational delay + Container position + Picking and dropping + Head movement

Fig. 8.5.1 Generalized timing formulae for magnetic devices

Devices can do some jobs independently, such as arm movement, tape back space and rewind, as only data movement ties up the channel. Channels are, in fact, a small computer in their own right with their own rather limited instruction set.

It is possible to have several input/output areas for one device so that you can read in record after record as a buffer against uneven processing time — normally about two areas per file. The device can carry out a sequence of instructions without any interruption of the main processing. Data is placed in core storage or taken from core storage in competition with the main processing program which also requires the storage mechanisms. This is done by 'cycle stealing' in which the main program stops for one core storage cycle and then continues. If this method is employed it is necessary to know how wide the data path is and the core cycle time. The required timing formula is then

$$\text{Time lost to main program} = \frac{\text{Total number of bits}}{\text{Width of data path}} \times \text{Cycle time.}$$

To calculate the overall run time, it is necessary to divide the run into components, time each component, summarize according to channels, processor and slow speed devices and then decide the longest times. These will form the run time. With experience the devices having a small effect on the run time can be timed by inspection and disregarded from the calculation, attention being devoted to the critical devices.

In some cases the reference to a file may be uneven. It is

Media	Formula
Paper tape punch or card punch	$\dfrac{\text{Characters punched}}{\text{Character rate}}$
Paper tape reader or card reader	$\dfrac{\text{Characters read}}{\text{Character rate}}$
Printer	$\dfrac{\text{Lines printed}}{\text{Print rate}} + \dfrac{\text{Lines skipped}}{\text{Skip speed}}$

Fig. 8.5.2 Generalized timing formulae for non-magnetic devices

necessary to know the *hit rate* (e.g. 5 per cent activity) and also the *hit grouping* (50 per cent of transactions refer to 2 per cent of the file). In order to time a run with an uneven hit rate, treat the run as two separate hit problems. Alternative systems approaches should be considered to avoid passing a tape file unnecessarily often, or having very large disc files on line continuously when only a small fraction of records or data in the records is required on a daily basis.

Reference to a disc file may be by one of the following methods: (1) direct addressing—imposing limitations on the use of reference numbers; (2) direct addressing by randomizing—use of a formula to convert actual references to direct addresses; (3) use of an index for each cylinder.

Use of an index will involve additional disc reading time whenever reference is made to a different cylinder. The time to read the first record on a different cyclinder would be given by the formula

Arm movement + Rotational delay + Reading index
+ Rotational delay + Reading record.

Reference to a record on the same cylinder would not normally require reading the index since this would presumably still be available in main storage.

When programs are sorted by the operating system on disc it may be possible to call in special subroutines from disc to cater for unusual conditions that arise during the program. These subroutines should be planned so that (1) they will all fit into a common transient area or areas; (2) execution of one subroutine can commence while the others are being brought in; (3) use of subroutines are the exception rather than the rule, otherwise the run time will be very long. Times can be calculated treating the disc as a normal input device.

In general many of the matters discussed in this section are relevant to program planning, which will normally be undertaken by a senior programmer rather than by the analyst. However, this brief outline has been given so that the analyst can discuss points of this type intelligently, if the programmer concerned suggests modifications to improve the timing performance of the system.

9 Justifying the System

Ensuring that the computer system justifies itself by achieving desired results is an area which is often forgotten. The only requirement in many cases is that the job runs somehow on the computer. Large sums of money are often spent without adequate return and inefficient personnel remain inefficient. Every job starts with a blank sheet of paper and a sharp pencil, and few standards of procedure of performance exist.

Obviously the desired results must be planned, with procedures and time schedules. It is necessary for those involved to understand standard procedures and the level of performance expected, but this does entail much more understanding of the underlying problems than is usually achieved. Many people are involved on the fringe of computer applications, but few find suitable training and even fewer are released from normal duties to get the time to achieve useful results.

The plan must have objectives and if these objectives are not met, adequate feedback to the plan and revision of objectives must take place. For example, if the systems analysis is behind schedule for one application, a thorough check must be made to see whether the factors causing this will affect other systems studies.

The success of using a computer is frequently the very fact that the organization's system objectives are met. Under the existing system, delays and errors may well mean a relatively ineffective system. The systems analyst will find many examples where people have learned to live with, and accept, inaccurate systems, simply because it is not their problem, or they just do not know how else it could be done. The speed, accuracy and checking techniques of the computer system should result in objectives being met, and met more accurately.

9.1 Defining computer benefits

There will always be people in any organization who will say that the system was much better before 'something' happened. When the 'something' referred to is the introduction of a computer, such people will revel in pointing out errors which have occurred as a direct result of computer processing. The analyst must be able to counter this scepticism with sound arguments of the benefits to be derived from the intelligent application of computers, and be sure that the system he is introducing will realize some or all of them.

The computer salesman's hint of staff saving is rarely the obvious justification for installing a computer system. In a few cases substantial saving in personnel is made, but in others, savings are offset by the need to recruit specialist staff who are more expensive. However, it is safer to say that once the computer system has the resources to function successfully, it is likely to be capable of absorbing more work through the growth of business without increases in staff.

A more significant benefit arising from computer systems is the more effective use of the organization's financial resources. For example, an efficient stock control system will require less capital and yet give more satisfactory results. Similarly, investment in a project may be more carefully controlled since resources can be purchased when actually required. It is not exceptional to find that a business has grown to the point where it may unwittingly be carrying along unprofitable lines merely because insufficient accurate data are available. This information may even come as a by-product of computer systems.

Growing organizations, that wish to continue to grow, use new equipment and new techniques, sometimes with no benefit but giving them an advantage over competitors. For example, many businesses bid for work by estimate and tender. In the past, a good estimating staff would know when they obtained a very profitable contract, but when the profit margin was relatively small, they could not estimate accurately and would in fact take some contracts at a profit and others at a loss. In addition, the many variables and their fluctuation meant inaccuracy, since a manual system could cover only the most frequent or most important variables. This is an area where the computer can shine, since any number of variables and complex estimating procedures can be calculated quickly and accurately; and changes calling for the recalculation of estimates can be serviced quickly.

The use of computers will deal a serious blow to 'hunch management'. It has been said that an acceptable business manager is one who makes more correct decisions than wrong decisions. If a computer can help to remove some of those wrong decisions for any manager, the business must increase considerably in effectiveness. It is usually the speed of provision of data by computer which is the greatest single factor in improving effectiveness through the management function.

To an informed observer it is often noticeable how slowly a business reacts to external changes. It is almost as if the business as a separate entity were determined to follow its original plans, and stifle all internal changes, unless the original plans are now actually impossible. The introduction of a computer system can sometimes bring about a remarkable psychological change through some such reasoning as 'the computer says this is now impossible—and we must act accordingly'. Substitute 'Jim Roberts' (or any other person) for 'the computer' and the same information is unlikely to seem so important. A fact given by a human being appears to be an opinion to others. Anything (it happens to be a computer) which highlights important exceptions and deviations quickly must markedly increase the effectiveness of a business.

There can be few systems analysts who have not noticed substantial benefits arising from the computer investigation itself. Even if the computer proposals were not implemented, these benefits would still be obtained. Most businesses 'grow like Topsy' and very few spare the time and money for a systems checkover unless a computer is contemplated. Of course this means that inefficiencies are built in and treated as part of the essential fabric of the system. For instance duplication of records will be found, reports will be produced which ceased to have any value years ago, and information will be rendered to managers much later than possible because of some impediment long since removed.

Most of the obvious cost-reducing systems, e.g. payroll, invoicing, general ledger, are now history. These systems were aimed at the operations area of the organization; they took over the 'donkey work'. Nowadays the emphasis is on information systems, i.e. systems which help the decision-making process. Systems have moved away from operations into the planning and control areas.

Consequently the justification of systems has changed. Direct savings are now taking second place to savings resulting from better management information. In particular there is a decreasing tendency to justify a system on the basis of reduced staff costs.

It is possible to identify a number of areas in which savings can be made and we can sum this up with the following checklist:

(1) *Direct savings — a justification for expenditure* Savings could occur in the following areas: cost of equipment replaced; reduction in clerical staff; reduced accommodation costs; reduced overtime charges; reduced communication costs.

(2) *Better management information* Speeding up the information flow and making it more accurate, comprehensive and meaningful can give: an indication of the most profitable product mix; less work-in-progress; fewer items in stock, especially slow-moving items; less manufacturing changeover, i.e. larger batches; an indication of those customers on whom the salesman's effort should be concentrated; better administration of credit facilities; fewer bad debts; less opportunities for fraud. None of these should be regarded as a benefit until someone in the organization puts a value on them.

(3) *Better planning* An effective stock control system may be able to reduce the costs of inventory, improve purchasing practices and raise the standard of customer service.

9.2 Cost assessment

At some stage an estimate has to be made of the overall cost of the new system. Unless this is done, senior management can have no firm basis upon which to compare the proposed system against any existing methods. In exercises of this type there will always be the intangible costs associated with 'better management information', 'faster customer service', and so on which are difficult to quantify in monetary terms. However, there are a surprising number of positive expenses associated with any computer installation, and the analyst must know the extent of these if he is to provide information as a basis for costing.

The first costs to be considered are operating costs of the existing system. There are likely to be manpower; materials; equipment; identifiable expenses (telephone, rent, electricity, etc.); working capital (inventories, accounts receivable, cash).

The proposed system will have additional operating costs: data preparation; computer; data control; system maintenance and modification; special stationery; any associated equipment. The proposed system will also have some significant development costs. These will include fact-finding, interviewing, etc.; presentations and reports; analysis; design; programming; testing; file conversion; parallel running; training. No generalizations can be made about the cost of staff, except to say that it is expensive and becoming more so. Most organizations will have a standard charge rate for various categories of staff. From this a daily or hourly rate can be calculated

Having established a suitable cost per hour for the systems and programming activities, it is necessary to determine the time required by this staff before the system can become operational. Such estimates will cover fact-finding, analysis, design and planning, program coding, and testing, plus an allowance for file conversion and implementation duties which the analyst will be involved in.

Fact-finding and analysis costs depend on the depth of the study, which can be anticipated only on the basis of enlightened guesswork. A number of analysts are assigned for a period to a project on the strength of a senior man's experience. Feedback is important to see how accurate the guesswork was. Accurate records must be kept of projects assigned and related costs.

In this difficult area, the practices of the better consultants are worth noting. Their systems frequently are costed reasonably accurately, sometimes because of extensive and varied experience, but also because the project's parameters and terms of reference are closely defined and amendments without written notice are not permitted. So often in fact-finding, the analyst is diverted to other areas, given wider terms of reference or delayed whilst awaiting management backing. In such cases original cost estimates are useless.

What has been said above about fact-finding applies also to design and planning. However, the computer manufacturers do a considerable amount of work in designing and planning, both during the selling operation and subsequently as user support. The inclusion of some of these items in the computer installation costs is often misleading. Much of this expense relates to an O & M/business organization overhead which, even if avoided so far, would eventually be necessary for the organization whether or not a computer was scheduled.

Costs of programming are difficult to estimate with any degree of accuracy at this stage. Algorithms based on program complexity, lines of code, etc., are not particularly helpful since approval for the system must be obtained long before these details are available. Many organizations now have a standard method of calculating development cost. The best the analyst can do is to make an informed guess based on the appropriate method.

The costs of converting existing files into the chosen computer medium is in the nature of a once-only charge which often occurs during the immediate post-installation period. The manufacturer can often give a good guide to the time involved which can be converted to cost. If the file is set up by an external agency, an estimate is normally available and, since performed by experts, may well be cheaper. However, the majority of installations use slack time in the early days of their computer for this task.

The analyst must also take into account any additional equipment his system may require, e.g. data preparation and related office equipment. Regarding the type and cost of such equipment, the analyst must be guided by the operations manager.

Running costs, other than for staff, include all the ancillary items necessary for the installation to become a working unit. Perhaps the major items in this group are the computer consumables, including stationery, discs and magnetic tapes. The absence of an adequate control on these ancillaries is often found in installations. Examples of waste are (1) using multi-part plain stationery for single-sheet listing jobs; (2) having special preprinted stationery for internal reports, which could be adequately produced on plain paper with computer-printed headings; (3) writing files on 2400 foot magnetic tapes when 600 foot ones would be sufficient for the contents. Moreover, these costs are frequently ignored in costing exercises, although they can involve expenditure of several thousands of pounds per annum.

The above are not intended to be exhaustive, but are suggested as guidelines only. Some miscellaneous charges for training courses, manuals, publications, and so on may be added. In any particular installation, the analyst should be able to draw up a comprehensive list as an aid to costing the system. How far he will be involved in this area of the work is uncertain, as it is sometimes a task undertaken by the data processing manager. However, even if he is not concerned with the overall costs, the analyst must be prepared to obtain costs for and justify those parts of the system which incur *additional* expense, such as special stationery, extra data preparation equipment and so on. In presenting this information it is useful if he understands the framework into which these costs must be inserted.

10 Presentations and Reports

During all stages of his work, the systems analyst will find a need to communicate his ideas and designs to others. Such communication may be formal or informal. Examples of formal methods are progress reports about his assignment or a completed Systems Specification for handing over to a programmer. Informal methods range from chats within the computer department with other analysts to 'off-the-record' discussions about the new system with operational staff in the user area. As already stated in Chapter 1, the analyst must possess communication skills of a high order, if he is to succeed in convincing others that his ideas are practical ones.

Communication, in this context, includes the need for competence both verbally and when preparing written matter. The analyst may be called upon to explain his system to audiences of widely differing backgrounds, from senior management to shop-floor workers. Similarly, his written skills must include the ability to prepare concise reports, and any other documentary support necessary for implementing his system. The sections which follow outline the main items the analyst needs to comprehend to perform his communication task effectively.

10.1 Communicating the system

The introduction of any new system will have far-reaching effects if it properly integrates work flows. It will affect the future work pattern in user departments, change their functions and often alter the responsibilities of the staff in them. Some existing departments may disappear altogether and new ones may be created. In such an atmosphere of change, it is useless to expect the system to work unless both its details and the method of implementation are properly communicated throughout the organization.

Much of this task can be handled through an implementation committee with members drawn from the departments concerned. If this is done the functions and scope of all communications must be defined, planned and controlled. The committee will ensure that the following areas are kept informed: (1) departments carrying out the changeover; (2) user departments directly affected—and this means *everybody* in the departments concerned; (3) senior and middle management who will be indirectly affected.

The total communication task will incorporate education, training, reporting and control. Education involves broadening people's minds. The introduction of a new system will produce disturbances of the daily task which make for insecurity. Even those not directly affected may have a feeling of unease over future extensions of computer systems. Some of these fears must be allayed before implementation and the analyst can help during interviewing and fact-finding to dispel the mystique of computers. During implementation, people should be involved by good dissemination of information on progress, showing the system in the context of the organization.

Ways of achieving this are (1) articles in staff magazines; (2) organizing visits to the computer centre; (3) discussion meetings where questions can be answered by the team implementing the work; (4) keeping all union representatives well informed.

Training is giving people new skills for old, teaching them how to do the job in the new way, even though their attitude of mind may not alter. It needs planning to ensure that the facilities for training are available and that the skills acquired are used as soon as possible. A comprehensive training program will require (1) handbooks and manuals; (2) lectures and demonstrations; (3) examples of new documentation; (4) visual aids to help in training; (5) good lecturing techniques.

Education and training are not one-way channels, as there must be a feedback. The system must learn and adapt itself, e.g. difficulties in teaching may reveal defects in form design. Feedback is the object of reporting. As implementation is being controlled there must be reports on progress.

Reporting needs to be formalized. If discussions are to take place, then they also must be conducted formally with minutes of agreement and disagreement and the action proposed. If action has to be carried out by another department, the completion of the task will have to be notified and a further report will be required. All this formal work must be controlled by the committee, who should also lay down rules for distribution of documents.

Problems of selection arise in distribution systems to ensure people get what they need. Keeping distribution lists, handling amendments to training documents and basic systems information are further communications jobs that the analyst must undertake. Reporting and control go together and there may be difficulties of establishing effective control cycles. If the committee only meets at irregular intervals, it may happen that the predetermined control cycles do not coincide with a committee meeting. Under these circumstances, the committee will not be able to review progress made and exercise an effective control. The solution to this problem is for the full committee to delegate its authority to one member whose task is to ensure that the required control function is executed at the proper time. The obvious candidate for the task is the systems analyst, and for this reason he will often find himself in the position of secretary of the implementation committee. As such he will be the centre point in the communications network stemming from his system. His aptitude for the job is obviously a vital factor in the overall success of the implementation.

10.2 Visual and aural methods of presentation

To express himself well is a necessity for the systems analyst presenting his new procedures for acceptance by management or the operational staff. A clear and concise exposition of the new system will do two things. It will help to overcome the natural resistance to change which exists in everybody and it will enable the new procedures to be implemented with the minimum of disorganization, because a fuller understanding will have been achieved.

Perhaps the most common way in which the analyst will communicate his ideas is by a formal lecture, normally to fairly small numbers of staff. The most important aspect of this task is adequate preparation. Remember that what is said about a subject should only be a small part of the speaker's total knowledge. It is a mistake to confuse the issue at hand by presenting too many arguments. Preparation is assisted if an overall framework exists into which the talk must fit. A suitable framework is as follows:

(1) State the proposal.
(2) Concede the objections.
(3) Support the proposal with *one* main argument.
(4) Give evidence to support the argument.
(5) Restate the proposal.
(6) Any questions?

A lecture can often be made more interesting, and points explained more clearly, if some kinds of visual aids are used. Various methods are available, and points to consider in using some of them are given below:

1. *Films*
   (i) Expensive to make
   (ii) Have to be chosen with care
   (iii) Usually of a general rather than a specific nature
   (iv) Very good as an introduction to the problems of EDP
   (v) Room has to be darkened.
2. *Overhead projectors*
   (i) Easier to produce transparencies than slides
   (ii) Sometimes too easy to produce and slip-shod results occur
   (iii) Can be built up by laying one transparency over another
   (iv) Can be written upon by Chinagraph pencil and then cleaned for re-use
   (v) Room does not need to be darkened.
3. *Chalk boards*
   (i) Ideas and diagrams can be built up and altered quickly
   (ii) The medium can suffer by being too flexible; insufficient thought is put into what is expressed on the board
   (iii) Very dusty in confined spaces
   (iv) Once cleaned it is not possible to refer back without redrawing
   (v) Not easily portable.
4. *Flannel boards*
   (i) Quicker to use than chalk board
   (ii) Diagrams can be built up piecemeal
   (iii) Very portable.
5. *Magnetic boards*
   (i) Quicker to use than chalk board
   (ii) Diagrams can be built up piecemeal
   (iii) Must have sheet steel as board
   (iv) Board can be used as chalk board also if suitably coated.
6. *Flip-charts*
   (i) Better than slides as room does not require darkening
   (ii) Easily transportable
   (iii) Can be drawn upon but not subsequently re-used
   (iv) Usually cheaper to produce than slides
   (v) Can suffer from slip-shod design.

The general principles to be observed when considering the use of any visual aid are as follows:

(1) Don't crowd the aid with too many facts.
(2) Don't put more than one main idea on one aid.
(3) Don't use aids for their own sake: have a purpose in mind.
(4) Remember it is more difficult to 'hold' an audience if the room is in darkness.

We have already stressed the importance of making sound presentations to management about proposed new systems. Perhaps this section should be concluded by a list of dos and don'ts concerned with presentation technique.

Do

(1) Come well prepared.
(2) Familiarize yourself with any equipment before the audience arrives.
(3) Introduce yourself and your colleagues.
(4) State the purpose of the meeting.
(5) Outline the presentation and give signposts to it.
(6) Concentrate on impact rather than details.
(7) Address your audience generally and specifically.
(8) Listen and be prepared to alter the order of events.
(9) Summarize your conclusions and end pleasantly.
(10) Leave time for questions.
(11) Have confidence in the system.
(12) Acknowledge assistance and guidance.

Don't

(1) Introduce surprises.
(2) Be flippant, casual or argumentative.
(3) Over-emphasize organizational weaknesses.
(4) Talk down to your audience or at your visual aids.
(5) Use computer jargon but plain English.
(6) Lose control of the meeting.
(7) Lose heart if things don't seem to be going entirely your way.

10.3 Report writing

Reports are formal communications of the reasons for, the nature of, the results of, and the conclusions from a particular course of action or investigation. A report is generally made by someone instructed to do so by a superior who has authority to make decisions relating to the subject matter of the report. There is an implicit obligation on the people calling for a report to define the use to which it will be put. There is an obligation on the writer to supply all the relevant information in an unbiased way; he always knows more about the subject of the report than his intended readers, otherwise they have no reason to ask him to write.

As reports are often intended to persuade people into initiating action, they should be particularly oriented towards the character of the recipient. This is not to say that the facts quoted need not be true or that opinions need not be honest, but that the report and language should be structured to impress and satisfy the intended reader for whom action is desired. Such colouring of a report is ethically justifiable only if the contents are true, valid and honest. It is not justifiable to state only partial truths or to be vague about things that can be precisely stated, e.g. to say 'a high percentage' instead of 51 per cent or 'a survey was made' when the writer talked to three out of 10 workers.

Since reports are intended to be objective, it is usual to write them in an impersonal style, although this may reduce the impact. This can be overcome to some extent by interspersing the report with short terse sentences, or even phrases standing on their own. Grammarians may object, but they are not the sole arbiters.

The subject matter and objectives of reports are very varied, but there are certain principles which are generally accepted as being reasonable. The introduction should state why the report is being written, the terms of reference of the activity being reported and any brief qualifying facts which are necessary for the understanding of what follows.

The main body of the report should be divided into sections which have a logical unity in the context of the subject being discussed. Another obvious rule, often ignored, is that there should be a logical sequence to the sections; this may, for instance, be the order in which the subject matter of the survey takes place. This sequence should lead the reader to reach the same conclusions as are stated in the report. In order to maintain interest, detailed procedures, specifications, charts, etc., should not interrupt the main trend of the report, but should form the subject of appendixes, with brief reference in the script.

Reports should be written in ordinary clear English. Attempting a formal style to which the writer is not accustomed will give results which are stilted, artificial and unconvincing. However, special attention must be given to clarity, and an aid to this is to realize that there are three main types of information being imparted:

(1) *Descriptive or factual information* This indicates the existence of something. It may also *describe* inferences which may be drawn from the facts, e.g. in a car test report, the results of the test are fact, and so also are the inferences which may be drawn, but if the *validity* of the inferences is discussed, then the information is going beyond the merely descriptive.

(2) *Instructional information* This is designed to show how to do something or what to do. It can range from edicts arbitrarily imposed to suggested advice supported with arguments to convince.

(3) *Dialectical information* This is the communication of opinions and ideas, based on logical inference from a series of definitions or observations, and the statement of reasons for these opinions.

Consideration should also be given to semantic information. This term may be used to describe the sort of information existing primarily inside the mind, and is the collection of associations which are taken as the 'meaning' of a word. Imagine a system in which all information is purely selective — all the words in the 'dictionary' used by the system are previously agreed. Communication consists merely of selecting words and transmitting them. However, this does not tell us how meanings were ascribed to the words in the first place; this is a process of association and inference and is ultimately based on experience. Thought must therefore be given to the meaning of words. Distinguish, but do not ignore, the emotional content of a word. Prefer words which refer to specific identifiable facts and ideas rather than abstract generalizations. Remember that words may have 'private' meanings to you and to others, that new words may not be universally understood, and that the meaning of words may change over time. If in doubt, consult a recent dictionary.

The following check-list gives a brief outline of items to consider in report writing:

(1) *Defining task — terms of reference*
    (i) What exactly is being written about?
    (ii) Why is the report needed?
    (iii) What effect may the report have?
    (iv) Who are the readers and what do they know already?
    (v) What is wanted — guidance or firm proposals?
    (vi) What work has been done previously?

182 Basic Systems Analysis

(vii) What previous reports have been made and what were the conclusions?

(2) *Information contained in the report*

(i) What period does the report cover—a month, a year?

(ii) How old is the information contained in the report?

(iii) What is the source of each item of information quoted?

(iv) What procedure will be followed to check the information quoted?

(3) *Method of preparation*

(i) Which person or group is responsible for the preparation and writing of the report?

(ii) How long will it take to prepare?

(iii) How is the data contained in the report compiled?

(iv) How many copies are to be prepared and to whom are they to go?

(4) *Effectiveness of report*

(i) What is the report designed to accomplish?

(ii) What use will each recipient make of the report for his own purposes?

(iii) Does the report meet all the requirements of the terms of reference?

10.4 Data Processing Department reports

The previous section considered basic principles applicable to the presentation of any written report. However, there are some special purpose reports which are generated by the data processing function, and, as such, require particular attention by the systems analyst. These are the formal documents which summarize and convey the results of the work of the department, and are prepared at three levels: (1) *senior management*, to give advice on overall implications and advantages of new computer applications, and on the general running and efficiency of the department; (2) *middle and lower management*, to indicate the effect on their departments of new and revised procedures; (3) *internal to Data Processing Department*, i.e. systems specifications.

Descriptions of agreed systems, for detailed implementation, and reports on departmental efficiency are considered elsewhere, as are the detailed systems specifications to programmers. The main concern of this section is with the reports to senior and lower management concerning the adoption and implementation of new systems. The involvement of the systems analyst in the preparation of these reports will depend on his seniority. As a junior he will have done much of the preliminary investigation and details of design but will not usually prepare the report. As a senior he may well write the major part of the report to middle and lower management. The reports to senior management may be largely the work and responsibility of the data processing manager, or chief systems analyst, or project leader in the case of a large installation.

Each system, and therefore the formal proposals recommending its adoption, should be viewed from the following aspects:

(1) *The management system*   A non-computer exposition of the system, relating it to the management environment in which it is to operate.

(2) *The computer system*   A computer-oriented discussion of the way in which the computer is used, with particular attention to file-structure of the system.

(3) *The human aspects*   A description of the human functions involved at all levels of participation in the system, with particular attention to the detailed design and use of input and output forms and devices.

(4) *The operational system*   A description of the operational aspects, including hardware, and paying particular attention to timing and efficiency.

(5) *Implementation*   An assessment of the quantity and quality of resources (machines and people), and their timely availability, necessary to meet the project schedule.

(6) *Costs*   A presentation of the estimated costings for all phases of the project.

(7) *Design process*   An account of the work of problem identification, problem solution and system design.

Whenever possible, the presentation should identify points at which the proposed design is sensitive, or at which special measures have been taken to avoid sensitivity and to produce a robust solution. Points of sensitivity (i.e. points of potential instability in the solution) may arise under each of the above headings.

Senior management will receive what is termed a *feasibility report*. This is a broad study of an application area with particular reference to the suitability of data processing techniques in the area. It also considers the cost of instituting data processing procedures, having regard to the existing workload (system, programming and machine) of the Data Processing Department, and the cost and impact of the acquisition of additional computing resources if required. These reports usually originate from manufacturers (generally as 'proposals' for equipment), consultants (particularly if the company has no Data Processing Department, or wishes an independent evaluation of manufacturer's proposals), data processing managers, particularly in smaller installations, or project leaders. If acceptable to higher management, the report may initiate the detailed systems study, which gives rise to the *systems proposal*.

The systems proposal is the most important document produced by the systems analyst in his function of 'salesman' for data processing in his organization. Its purpose is to obtain management approval by convincing management that the system proposed will meet all the major requirements in an economic way. Once approved at this level, the detailed specification to programmers may be completed. The report should (1) assure management that the system will give them what they want; (2) show some benefit, usually financial—adequate work must therefore have been done to validate the accuracy of all the figures quoted, particularly file sizes and machine loadings; (3) show the system to be stable, in terms of changes in volumes and staff; (4) persuade management of the importance of their cooperation, e.g. in the provision of accurate input to the system; (5) emphasize the need to reduce to a minimum any changes to the system, particularly to outputs, during implementation.

The different purposes, the different depth of study and the different levels of management to whom proposals may be addressed mean that the amount of detail included in different reports will vary. A proposal to top management would concentrate on costs and benefits with little detail of the proposed system. A proposal to user management would specify in detail in the appendixes all aspects of the system relevant to the functions under their control. Therefore it is suggested that the management proposal outline shown below should be regarded as a check-list and not as a rigid standard. The information given must be the minimum necessary to enable management to make an informed decision. Any detail other than in the appendixes will be more of input or output rather than of processing. Features essential to all management proposals are the effects and implications of the new system, its anticipated costs, savings and other benefits.

A suggested format for a management proposal (divided into five sections) is shown on the next two pages.

1. *Preliminary matter*

Title page	Report title and system or report reference
	Author and department
	Month and year of publication
Contents list	Main headings and subheadings with section and sheet numbers
Summary	Objectives
	Proposals
	Costs
	Benefits
	The summary should not extend beyond one typewritten page
Introduction	Background to the study and reference to previous reports

2. *Terms of reference*

Of study	Scope, objectives and constraints as originally defined together with modifications made during the study
Of existing system	Relevant information on the organization and its development
	Outline of existing system
	Existing and anticipated problem areas
Of system requirements	Design requirements and constraints of a new system, rules governing operations, accuracy, quality, cost, future projections
Of proposed system	Outline of system
	Alternatives considered and rejected, reasons for rejection
	Implications of the new system of interest to management
	Organization and staff
	Equipment and software

3. *Development and implementation plans*

Costs	Equipment and software costs
	Systems development and implementation costs
	Operating costs in comparison with existing system

Benefits	Savings on current costs
(quantified	More effective utilization of resources
where	Information quality
possible)	Better control

4. *Recommendations*

5. *Appendixes*   Existing systems supporting information
Proposed system supporting detail
Hardware and software evaluation and selection
Implementation supporting information
Glossary

11 Implementation

The implementation phase of a project covers the period from the acceptance of the tested design to its satisfactory operation, supported by the appropriate user and operations manual. It is a major operation across the whole organizational structure and requires a great deal of planning. Planning for implementation must begin from the initial conception of the project. It requires a thorough knowledge of the new system — its personnel needs, hardware and software requirements, file and procedure conversion activities — and of the current system — where it interfaces with the new, the changes to it, the jobs which will be superseded, etc. Only the analyst responsible for creating the new system will possess this knowledge. The systems analyst can plan, schedule and co-ordinate but has no executive powers.

Planning must cover the following aspects: organization for implementation; control of resources; motivation of the users; training and the production of manuals; file conversion; changeover. These form the subject matter of the individual sections of this chapter.

11.1 The management of implementation

Implementation is the practical job of putting a theoretical design into practice. It may involve the complete implementation of a computer complex or the introduction of one small subsystem. While proving the design (*see* Chapter 8) many members of the staff were involved, not just the computer department; during implementation even more will be. The usual method to plan and control resources — manpower, computer time, etc. — is to appoint a co-ordinating committee.

As chairman it is best to have the line manager of the department most affected by the system since he has the most vested interests in it. The committee should include the data processing manager, representatives of all other user departments affected, a representa-

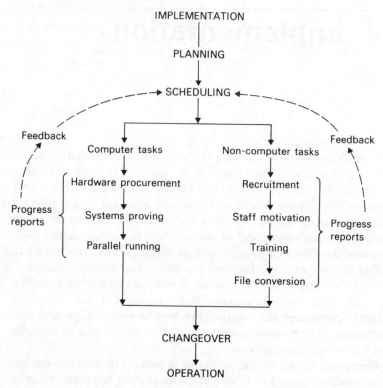

Fig. 11.1.1 Organization features in implementing a system

tive of the accounts department and the system analyst, who may well act as its secretary. In practice, much of the work and authority of the committee may be delegated to the systems analyst since control by committee is usually fraught with delay, indecisiveness and frustration. If the committee is to function effectively, its members must work as a team. Responsibility for supervising tasks must be delegated. Regular meetings to review progress, adjust schedules, air ideas and plan new steps must be held to optimize the use of resources.

Figure 11.1.1 shows the main organization features involved in implementing a system.

11.2 Scheduling the tasks

In the previous section the importance of planning was stressed.

Planning is deciding in advance what is to be achieved and how it will be done. This necessitates rational thinking, and precision in defining action and results. Planning includes forecasting. Facts must be projected into the future and the basis for projection must be sound. Planning needs statements relating time to what is to be achieved.

There are two common aids to planning. These are charts and networks. The most usual type of chart is the horizontal bar chart, which is sometimes called a Gantt chart. Figure 11.2.1 illustrates a typical bar chart. In it, activities are listed in the left-hand column, and a time-scale is drawn across the top. Squared paper makes for easy construction of such a chart, and facilitates reference to the individual entries. Against each activity, an open oblong is inserted in that part of the relevant row to indicate the estimated length of time the job will take. The start of the oblong is positioned below the date on the time-scale which corresponds to the planned starting date of the job. The oblongs are shaded in to indicate time spent on each task. A cursor, which might be a transparent ruler or a piece of cord, is placed in the current date position and it can be easily seen from the unshaded portions of the bars which jobs are behind schedule.

The basic disadvantage of charts lies in their inability to show clearly and unambiguously interrelationships and the dependence of one task on another. Networks overcome this disadvantage as their function is to show interrelationships, although the time-scale is

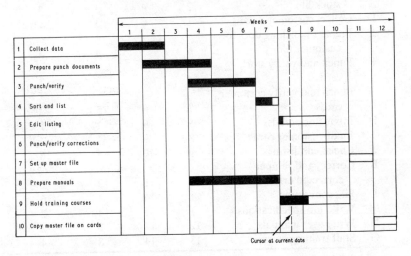

Fig. 11.2.1   Bar chart

abandoned. A mixed system may be used with a network for planning scheduling and overall control, and charts and graphs for control of small sections of the task.

Figure 11.2.2 illustrates a typical network diagram. Before considering how such a diagram is constructed, the following definitions are necessary:

(1) *Activity* — the application of time and resources that are needed to progress from one event to the next; spans a time period.

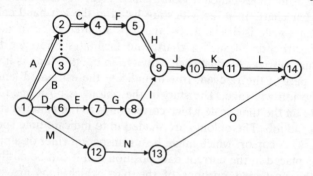

Event	Description	Head	Tail	Duration (days)	Manpower required
A	Collect stock item details	1	2	4	2
B	Collect stock price details	1	3	3	2
C	Prepare punching docs for stock file	2	4	3	4
D	Collect customer details	1	6	3	2
E	Prepare punching docs for customer file	6	7	4	3
F	Punch and verify stock cards	4	5	7	2
G	Punch and verify customer cards	7	8	5	2
H	Input stock cards	5	9	1	1
I	Input customer cards	8	9	1	1
J	Check edit listings	9	10	5	6
K	Correct and reinput corrected errors	10	11	3	2
L	File conversion	11	14	1	6
M	Write job specifications	1	12	4	2
N	Write user manuals	12	13	7	2
O	Staff training courses	13	14	8	4

Fig. 11.2.2 Network diagram (critical path indicated by ⇒)

(2) *Event*—a specified objective in the overall plan to be achieved at a particular instant in time.

(3) *Critical path*—the path traced through those activities in the network which together constitute the longest overall time.

(4) *Float*—the excess time which can be added to any activities not on the critical path, without altering the overall completion date of the project.

To draw up a network diagram, the first step is to list all the events in the plan, and number these sequentially for reference. Against each event, note the numbers of events immediately preceding and following it, and a realistic estimate of the time required for the particular activity involved. Plot the events on the diagram by numbered circles and join the relevant circles by arrows. The length of the arrows has no significance. Enter the activity times on the diagram against each arrow. Now, by tracing every possible path through the diagram from start to finish, a total time estimate can be established for each one. The path with the longest time is the critical path, and has no float. All other paths contain a float, which is equal to the difference between their overall time estimate and that for the critical path. A start and finish date can now be established for the overall project and for all its component pieces. Provided events on the critical path are completed on schedule and any delays on non-critical paths do not exceed their available float, the project will be completed on time. This fact enables a tight monitoring system to be established over the progress of the whole plan, and is one of the benefits derived from network planning.

When the number of events and activities in a particular plan is small, analysis of the network is feasible by hand computation. With a network of any size, however, the evaluation of the alternatives is tedious and error-prone. In the latter case, it is advisable to use a computer to perform the necessary calculations and many standard application packages—usually called PERT, meaning program evaluation and review technique—exist for this purpose. Network planning, as a general management technique for project control, has reached an advanced state since its introduction in 1957, and the analyst is referred to the many standard texts on the subject for a fuller appraisal of its possibilities.

The advantages offered by networks are as follows:

(1) Each job must be examined and defined in great detail to create discrete tasks.

(2) Networks show relationships which are an aid to planning.

(3) They put a time limit on the overall task, thus assisting scheduling.

(4) They help the distribution of resources, and consequently allocation.

(5) The measurement of achievement against the plan is assisted by the assessment of impact on forward tasks and control.

Disadvantages of networks are that:

(1) They are not drawn to scale.
(2) The chart *itself* cannot record progress.
(3) The current position is not immediately obvious.

Once the network is constructed and the critical path has been established it is a simple task to convert it into a conventional bar chart to be used for progress control. This is done by listing all the activities shown as individual bars on the chart. The activity time will determine the length of the bar, and entry numbers will appear below each other on consecutive rows. If resources available are converted to a similar time-scale, these can be superimposed on the basic chart and any float time available can be used to adjust activities to resources. The completed chart, combining activities and resources, can then be used to plan recruitment and training.

The daily manpower requirement has been plotted in Figure 11.2.3. It fluctuates wildly; staff allocated to the analyst should be as few as necessary and all should be fully employed. By making the following adjustments it is possible to stabilize the manpower requirement considerably (*see* Figure 11.2.4):

(1) Allocate 4 men to event A.
(2) Allocate 3 men to event B.
(3) Delay start of event D by 2 days.
(4) Delay start of event I by 2 days.
(5) Split event O into two 4 day periods.

The following list summarizes the points to remember in planning the implementation; they apply regardless of the scheduling method employed

(1) Think precisely and reduce overall implementation to a series of small manageable tasks for which you can 'guesstimate' a time target.

(2) Check your intermediate target dates. Are you in front or behind? Will this affect the final target date?

No.	Event Limits	Duration
A	1,2	4
B	1,3	3
C	2,4	3
D	1,6	3
E	6,7	4
F	4,5	7
G	7,8	5
H	5,9	1
I	8,9	1
J	9,10	5
K	10,11	3
L	11,14	1
M	1,12	4
N	12,13	7
O	13,14	8

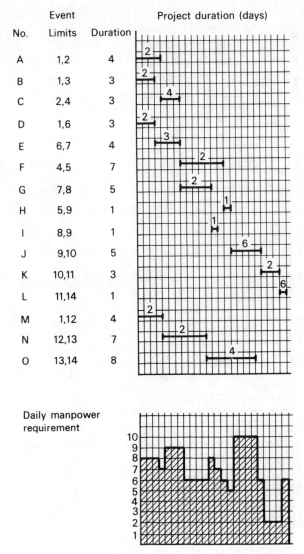

Fig. 11.2.3 Resource scheduling 1: events specified in Fig. 11.2.2

(3) Avoid continually revising your dates, or people will think you can't plan at all!

(4) Make sure everyone is involved in planning your targets and in doing the work to meet them.

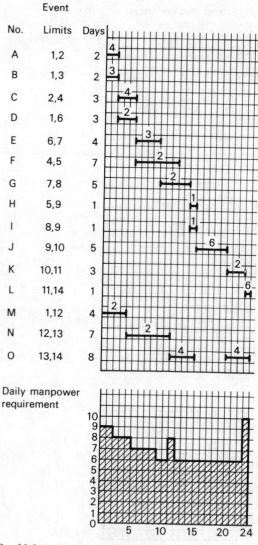

Fig. 11.2.4 Resource scheduling 2

11.3 Motivation of the user

Although senior management usually initiate a project by calling in a systems analyst, they will need to be motivated to accept the new system in much the same way as the lower levels of management.

For management the motivation should consist of (1) interviews and consultations to determine the requirements and involve management in the overall planning and scheduling of tasks; (2) presentations to explain the system and gain its acceptance; (3) briefs, demonstrations and conferences to assist in equipment selection and procurement.

At all levels of management and staff you will meet resistance to the introduction of your new system. In Chapter 1 the reasons for people being resistant to change were explained fully. You must remember that the system which has been designed will involve people and that their fears, suspicions, resentment and ignorance must be overcome by carefully planned meetings in order to (1) explain the purpose of the investigation; (2) give people the opportunity to participate and make suggestions; (3) acknowledge suggestions, to show how they have been incorporated, or explain why they have not been incorporated, in the design; (4) build up an awareness of the benefits of a computerized system so that their resistance to change will be converted to a desire for it; (5) explain the time-table for implementing the new system and the effects of the new system on each individual.

Although the systems analyst should be involved with motivating management, because of the time and effort involved line managers and supervisors will have to organize and chair staff meetings within their departments.

11.4 Training, handbooks and job aids

Resource allocation, as shown on the networks and charts, will decide the need for staff at various times in the implementation cycle. Realistic estimates must be made in this area and due consideration given to the problems of recruitment and training. Staff requirements will vary between the creation and changeover periods when the system is being prepared for operation, and will vary again at the operation stage.

Each job in the system must be specified. Certain jobs will have job specifications already written in the old system which will be unchanged in the new. Other jobs will require completely new job specifications to be written. These should include (1) main tasks involved, indicating relative importance of each; (2) the appointment to which the appointee is responsible; (3) supervisory responsibilities; (4) financial responsibilities; (5) discretionary powers;

(6) level of confidentiality of information handled; (7) type and valuation of equipment/material handled; (8) group relationships involved; (9) communication tasks, internal and external to the firm, indicating the frequency and method of communication; (10) working conditions, i.e. physical (noise, heat, humidity, etc.), psychological (stress, shift work, etc.), travel requirements; (11) job career prospects, i.e. the place of this job within a planned career structure.

In the final phase, it is possible that some existing staff will be transferred from departments whose function has been superseded by the new system. For example, ledger clerks may become control staff in a computerized invoicing system maintaining the sales ledger on magnetic tape.

When the attempt to match available staff with new jobs fails to complete the recruitment process, and some jobs remain vacant, there is a need for decisions to be made. Can existing staff be retrained? To answer such questions, the analyst/personnel team must consider the probable effects of delay caused by training and also assess the effectiveness likely to be achieved after the training is considered. The advice and experience of the personnel department will be of great assistance in the latter situation.

The cost of advertising, selection and training are extremely high. Recruitment delay can be reduced by specifying your requirements to a recruitment agency. Recruitment is essentially the responsibility of the personnel department; however, since data processing is a specialized area, the systems analyst should give the personnel department as much advice as he can to ensure that the right person is selected for the right job. In addition to the job specification the systems analyst should prepare a personnel specification. This should be based on the job specification and on the likely background, qualifications and abilities which an applicant will require to do the job successfully. The personnel specification might, but need not necessarily, include the following:

(1) Physical characteristics
    (a) height/strength
    (b) hearing, eyesight/colour recognition, speech
    (c) appearance/dress
(2) Qualifications
    (a) courses passed
    (b) programming languages used
    (c) languages spoken—if required to travel or work in foreign countries, or with foreigners

(d) previous job responsibilities and experience
(e) educational qualifications—graduate, 'A' levels, etc.
(f) special aptitudes—verbal, manual, mechanical or numerical
(g) driving/piloting ability
(3) Interests—intellectual, physical and social—which may help the selection of the most suitable applicant.
(4) Disposition and personal circumstances
    (a) type of character required (e.g. go-ahead, commanding, dependable, able to fit in well, etc.)
    (b) age group or background
    (c) domestic considerations (mobility of the job)
(5) Any factors which would automatically exclude an applicant

From these two specifications the personnel department should be able to decide the best way of attracting acceptable staff and the method of selecting the most suitable applicant.

If training is found to be necessary, proper schedules must be set up for this. To establish such schedules, due recognition must be given to the time required to arrange courses. Once a course has been planned, arrangements are necessary to release personnel from their current jobs to attend, and to be able to continue training after the course is complete under practical conditions. Obviously, the trained personnel will be needed as soon as implementation begins, and this implies that the planning of the training activity must commence a considerable time before this. It is often advisable to appoint a member of the implementation committee drawn from the personnel department to ensure that these factors are given proper consideration. Also, if the task is to be successful, the committee must have authority to arrange for the release of staff, when these are to be recruited internally.

It is important to set standards to measure effectiveness to staff after training and to prepare for the possibility of failures. Reports from independent assessors will help in this. Remember that monitoring on-the-job training is difficult, as reports can be subjective, but for personnel involved in file conversion, measurement of achievement is possible. Appraisal of post-training reports should be approached from point of view that, if a failure has occurred, the following questions are relevant:

(1) Is it the system's fault; has operation revealed a defect or weakness?
(2) Is the job right; can it be altered?

(3) Is the trainee wrong; if so is it a defect in training, or, as a last resort, a failure in the man?

With this approach one can ensure that if staff changes are made, correct replacements can be substituted.

During creation and changeover new methods are being introduced and these must be defined clearly so that men and machines can be told precisely what to do. Programming tells the machine and, similarly, the system must tell people: all the instructions must be laid down. Whereas computer instructions have to be represented each time, as there is no retentive memory, humans learn and remember. But the memory is fallible and it is sometimes more effective to present a guide to the job to aid the memory.

This gives a basis for defining (1) the *handbook*, which contains a detailed description of how the job is to be done; (2) the *job aid*, which is designed to assist a person to carry out the correct instructions whilst *performing* the job. Both are means of communication with different functions.

Within a data processing system, the following types of handbook will be found; the functions shown in brackets indicate the persons normally responsible for their creation: (1) computer operating manuals (programmers); (2) software manuals (manufacturers); (3) systems specifications (systems analysts); (4) program specifications (chief/senior programmer); (5) punching instructions (operating manager/systems analysts); (6) clerical procedure manuals (systems analysts.)

All these various types of manual are aimed at different audiences, and the level of presentation within each will vary accordingly. With some handbooks, the design should be arranged so that specific sections can be abstracted for issue to affected staff. This type of arrangement requires that sections are self-contained and do not include extensive cross-reference.

Standardization of manuals is impossible due to the disparity of their contents, but is necessary within related areas, e.g. software and program specifications. Certain standards can be maintained, however, such as using consistent paper sizes, binder colours and separator sheets.

Apart from the handbook, the job aid can be of great assistance in ensuring smooth implementation of the new system. Job aids can take innumerable forms of which those listed below are just random examples:

(1) *Form design*—aids to completion with headings; emphasizing points by boxes; colour, '3-D' printing.

(2) *Wall charts*—to show the scheme of processing runs; large diagrams of forms.

(3) *Notices*—placed in a position where they can be related to task, e.g. switch positions.

(4) *Colours*—to identify pieces of equipment; pipes and cables; coloured warning lights and status lights.

(5) *Colour in data preparation*—various paper tape colours to distinguish between verified data, raw data, control messages, programs, and computer output; cards edged with colours for identifying types.

(6) *Colour in result distribution*—coloured forms with matching coloured distribution envelopes.

(7) *Miscellaneous*—flow charts and decision tables for error handling, which are easier to use than a handbook; operating instructions for programs with specific actions required at each stage.

All the above, both handbooks and job aids, are aids to communication. In their design, always consider *with whom* you are communicating, *why* you are communicating and *when* you will be communicating. These three points will prevent the following fictitious example ever becoming reality.

A bell rings in a computer room. The operator looks up under Operating Instructions 'bell' and reads

'If ringing emanates from a bell painted red, mounted on the wall to the rear of the operator when facing the console, this indicates that the temperature and/or smoke alarms have been activated. These alarms will only be activated when the temperature is 90 per cent and/or smoke density is 1–2 per cent of normal. In such event refer to Emergency Manual EM 693/5, latest edition, Sect. 7, para 5(b).'

Finding the manual and the section the operator then reads

'The activation of the alarm system as described in EM 693/5, Sect. 3—BELL indicates the presence of a fire. In order to extinguish this and to prevent the spread of the fire, one minute after the alarms are activated and the bell commences to ring the automatic carbon dioxide system will come into operation and after a further two minutes (during which time the bell will keep ringing), the doors to the computer room, maintenance area and paper store, all within the area protected by the carbon dioxide system, will be automatically locked. Anyone who has to remain in the area will have to wear breathing apparatus (for location of such apparatus and for method of use see EM Sect. 5 para 3 and 9), but it is recommended that no personnel remain in the protected areas and they should be evacuated as soon as possible after the bell has started to ring. The locked doors cannot be opened until the carbon dioxide concentration has been reduced by use of the air conditioning apparatus (for details *see AC Operating Manual* Vol. 2, Sect. 8), and has been tested by switching on the special detector (*see* OC Vol. 1, Sect. 23, para 6 for full instructions).'

A 'job aid' of flashing lights bearing the word 'FIRE' and perhaps a reminder of the time element would have been preferable!

11.5 File conversion

File conversion is a system in its own right and will involve all the problems of fact-finding, data capture, clerical procedure design, form design and program specification. The way in which the job is done depends upon the size and complexity of the files and the old and new recording media, which may involve (1) manual systems of card index, folders, or binders; (2) machine systems of punched cards, disc, or tape; (3) records may be in master files, or small specialized files; (4) centralized or scattered filing systems.

The result of the conversion job is the final file as specified in the program. Each field on the file will be identified with a source document and sometimes information from a number of sources must be drawn together to make up the file. The ideal solution to the problem would be to be able to go directly from the existing file to the new file, but this is usually impossible and a number of stages are required. The ideal can be achieved only if the file to be converted already exists on a medium which is also acceptable as computer input, and this is rare. One notable exception is when existing files are maintained on punch cards but, even in this case, direct conversion is influenced by the quality of the existing cards, the code structure used on them and whether they are truly acceptable input. For example, a punch card file which had been maintained on 40-column cards could not be used for input to a computer having a standard card reader, without an intermediate conversion process. Similarly, existing magnetic tape files must conform in respect of size, layout, codes, labels, etc., before a direct transfer can take place, and so one installation may not be able to pass its files directly to another one which has a different computer.

As a result file conversion normally develops into the following sequence of events:

(1) Record the old file data on specially designed input documents by clerical effort.

(2) Transcribe the completed documents to suitable media, and verify.

(3) Use a tailor-made program to read transcribed data, and then output the required files in the format needed by the user program.

One of the problems encountered is that most source documents contain historical data which are not relevant to the new computer file, and this gives rise to the need for the clerical activity mentioned above. This in turn creates a further problem of the need for technical knowledge by the clerks involved, and often requires the release of experienced staff from operating departments. Sometimes such people cannot be spared for long periods, and the analyst must resolve any conflicts which arise from the demands of the old and new systems. It is sometimes possible to enlist the services of outside bureaux, or data preparation services to assist in the clerical and transcription-to-input media efforts. This may be a viable proposition when the technical content of the work is small, on the clerical side, or existing data preparation resources are inadequate for a large conversion job.

The greatest problem is that of converting a live file, such as a stock file. It must be converted at a specific date or series of dates and thereafter an amendment correction procedure designed. All this involves additional clerical work, and a decision has to be made as to when updating will take place: either during the creation stage or on completion of the new file.

A final task which the analyst must organize is how the new files are to be created. The options available are to do the whole job in one major effort or to take on parts of the old records at intervals and then amalgamate them. The choice depends upon the size of the task and whether time is available to prepare the required amalgamation program. In all cases, a special purpose program is likely to be needed to take on the new file, and this must be documented with as much care as is needed for any operational program. If the file is going to require a file amendment program when used within the system, thought may be given to whether a special take-on program is needed as well. The file could be created by treating all input records as insertions, and using the standard file amendment program for the task.

Eventually, the files will be set up, and the analyst must satisfy himself that they are accurate. He must consider that they will be no more accurate than the originals, and probably less accurate, as each stage in file conversion gives a chance for errors. The analyst must ask: Are the data in the source file correct? Are the staff available to check source records? What can they be checked against? Is there any chance of getting control totals?

This then raises the problem of controlling rejections to ensure that all records are converted. Continuing controls must be maintained until final reconciliation, although it is not always possible to guarantee the accuracy of take-over, because (1) it is difficult to correlate data; (2) a manual file may be active during take-over, raising the problem of what is the best time to take over — holidays, weekends, etc.?

The analyst must remember how essential it is to have some way of checking new files. The user departments must have confidence in the new files. While the files were in their departments they were always reliable, but once they are in the computer department it is amazing how wrong they can get!

11.6 Changeover

When the new system is fully tested and proved, changeover from the old to the new is possible. This can be accomplished in one of three ways, by (1) direct changeover, (2) parallel running, (3) pilot operation.

Direct changeover implies the introduction of a completely new system without any reference to any previous similar systems which may exist. The method is normally adopted only when there is insufficient similarity between the old and the new systems to make parallel or pilot runs meaningful or sometimes when the extra staff required to supervise parallel runs are not available. It is usual to introduce a new system directly at a time when work is slack, so that affected personnel can become accustomed to the change without extreme conflicting demands upon their time. It is essential, if a direct changeover is to be adopted for implementation, for the analyst to have previously established complete confidence in his new system.

Parallel running means processing current data simultaneously by both old and new systems in order to cross-check the results. The objectives of parallel running must be closely defined and a time limit set. If the method is to be considered as an extension of the testing for the new system, this is only useful if the old and new outputs are strictly comparable, which is rarely the case. The difficulties of cross-checking must also be borne in mind. If a difference is found, a decision must be made as to which system, the old or the new, is wrong; there is a natural desire to condemn the new system and any errors foster this spirit.

Another reason for using parallel running for implementation is as a standby in case the new system breaks down, as no irrecoverable position will have been reached. When this is the reason, it is important to set a firm time limit on the number of cycles for which the two systems will operate, remembering the cost and difficulties of running two systems and the possible effects on other computer work.

Pilot operation can be of two kinds:

(1) *Retrospective parallel running* is a procedure wherein the new system is run using input data from a previous period so that the results are known and can be checked. It does not simulate day to day running since it is not so time critical, but it does use less resources than parallel running.

(2) *Staged changeover* (otherwise known as phased changeover or restricted data running) is a method in which the new system is introduced in a piecemeal fashion, by phasing in different areas of work gradually. This makes personnel transfer easier.

Certain matters require the analyst's attention during this phase of his job. These are as follows: (1) co-ordination of changeover; (2) communication throughout the system during changeover; (3) controlling errors which may appear (i.e. how much system change and program modification are you prepared to allow at this stage?); (3) remembering that the change affects people; (4) testing is not complete yet: monitor the system and get feedback; (5) system maintenance methods must be designed and working, so that modifications can be effected quickly and successfully.

12 Hardware

A full discussion of computer hardware would require a book in its own right. Each manufacturer has his own ideas on the best design characteristics and computers made by different companies are not strictly comparable. New machines and peripherals appear so frequently that it is impossible to possess a thorough knowledge of all current hardware.

The trainee systems analyst need only concern himself with the very basic details of central processors and peripheral equipment. He is unlikely to become very involved with their finer design points and need only know enough to be able to follow any discussion of the topic in which he is present. The manufacturers will supply detailed technical literature for the analyst who requires more information on the capabilities of a particular model. The remaining sections of this chapter describe equipment with which the analyst is likely to become more closely associated.

12.1 Central processors

Central processors exist in mainframe computers, in minicomputers and in microcomputers. The exact definition of each of these is a matter of debate but the following is a reasonable guide:

Mainframe A computer of sufficient power so that it may be used by many people simultaneously for many different purposes. It may be concurrently executing an on-line database system for stock control, a time sharing system to support programmers in program development and a batch payroll system. It will support a very wide range of peripheral devices including disk drives, tape units and local and remote input and output terminals of many differing types. Mainframe computers are usually located in purpose built computer rooms.

Minicomputer A physically smaller computer that will typically be located in a normal office environment. It will usually support a number of concurrent users but they will not be accessing the varied range of systems software that is usually available to mainframe users. It will support similar peripheral devices to a mainframe but they will usually work at slower speeds and will be fewer in number.

Microcomputer   A small desktop machine which will usually be used by one person at a time to do one thing at a time. It will have integral diskette storage and also perhaps a hard disk and perhaps a tape device for taking copies of the hard disk.

There are more sophisticated micros which can run different software packages concurrently and can support a limited number of users with separate work stations—either VDU terminals or other micros.

A mainframe computer will contain one or more central processors plus other processors which each have a specific task to perform. Figure 12.1.1 shows the internal architecture of a typical mainframe computer.

The central processor consists primarily of the Instruction Element and the Execution Element. The Instruction Element is responsible for fetching program instructions and data from main storage in the correct sequence and preparing instructions for ex-

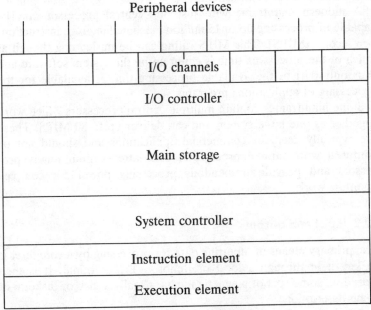

Peripheral devices

| | | | I/O channels | | | |

I/O controller

Main storage

System controller

Instruction element

Execution element

Fig 12.1.1   Typical mainframe internal architecture

ecution. The instructions are then fed to the Execution Element (sometimes known as the Arithmetic/Logic Unit) to be processed.

The system controller co-ordinates the activities of the other units of the computer and directs their operation.

Main storage consists of silicon chips which house the programs and data currently in use. It is common to find ten or more different programs in main storage at the same time on a large mainframe. Although main storage may be hundreds of megabytes in size, it still cannot house complete data files being used by active programs. Thus, the data is typically brought in to main storage from a peripheral device a block at a time, as and when it is needed. Output data is collected in main storage until there is a complete block of it and then it can be moved out to an appropriate peripheral device.

The Input/Output Controller handles this movement of data between the peripheral devices and main storage. This data is moved through I/O channels which check the validity of each byte and count the number of bytes transferred to ensure that no data is lost. Each channel can have a number of peripheral devices attached to it and these may be of different types. One channel may have an entire network of terminals attached to it whilst another may have just a few disk drives attached.

A modern mainframe with just one central processor may be capable of processing up to 15 million machine language instructions per second (MIPS). This MIPS rating can be misleading though as much of the processors time is eaten up by the system software and thus only part of the available processing time is available for the processing of applications programs.

Some mainframes contain multiple central processors which work together as one large system and can deliver up to 80 MIPS. These are typically used for commercial applications and should not be confused with parallel processors. The latter contain many processors and provide tremendous processing power for complex scientific work.

12.2 Input and output

The primary means of inputting data for processing by a computer is to key it in through a local or remote VDU terminal It is then stored on some sort of magnetic media, usually a disk or diskette of some description.

Punched cards are rarely used now and many computer installations no longer have card reading/punching devices. Cards took up too much physical space, were slow to read into the computer, were prone to damage and required much manual labour.

The main advantage of using VDU terminals is that they are connected to a computer of some description and can thus use the processor for data validation. The user is prompted on the screen to enter the appropriate data in the correct format and sequence. As each item of data is entered, the processor can check it for validity. In the event of an obvious error, such as an alphabetical character in a numeric field, the user is informed that an error has been made and is then prompted to re-enter the data. However, this does not provide protection against all data entry errors and thus verification may also be required. This can be accomplished by another operator keying the same data through another terminal. The processor can then retrieve the appropriate record from the magnetic disk and compare the two sets. Discrepancies will cause a warning to be transmitted to the operator, thus allowing correction. The warning may be in the form of the keyboard locking or of a visual alarm on the VDU screen. Corrected data is then written back to the magnetic disk.

Once complete batches of data have been keyed and verified, the supervisor can initiate the transfer of data from the magnetic disc for input to the main computer. This is usually achieved by dumping on to magnetic tape, which is a more compatible transfer medium. Some key-to-disc systems are communications-oriented and can transmit the validated data direct to the computer.

Key-to-diskette and key-to-cassette systems are free-standing, single-operator key-stations for the preparation of data on diskette and magnetic tape cassette respectively. Both systems consist of a microcomputer, a keyboard device (usually a VDU but typewriters are common with key-to-diskette systems) and appropriate media recording units. Data is keyed by the operator from source documents and is then validated and recorded on the storage medium.

With key-to-diskette units, the data diskettes are removed and then transferred to the main computer system. Data on magnetic tape cassettes is usually transmitted to another terminal or to a mainframe computer over communication lines.

Data collection systems are to be found mainly where data is not readily available in source document form but exists in a coded form, usually at dispersed locations. Typical applications include factory systems collecting production data and retail systems collecting sales data.

Whereas data preparation and input was once the work of specialist operators it is now often undertaken by end users who may be payroll clerks, accounts clerks, travel agents, shop assistants or members of the public. Anyone using a plastic card to withdraw money from a cash dispenser is acting as a temporary terminal operator. In such systems there is no requirement for intermediate data preparation staff although security and data integrity is of crucial importance. Without data verification by an input data operator there may be a higher incidence of keying errors but the cost considerations often make this acceptable. Thus, all transactions must be recorded by the computer for subsequent checking if it should prove necessary.

Where it is not feasible to use a VDU terminal, a specialist data capture device will be used. In the retail trade, staff might use a bar code reading wand, together with a keypad and a cassette recorder to enable them to walk around the store and check stock levels. The wand can be passed over the bar code for a particular item and the number of items can be visually counted, keyed in via the keypad and stored on the cassette. The contents of the cassettes can subsequently be transferred to disk for processing.

Although much output is now routed to VDU terminals for immediate inspection by the user, there is still a large requirement for hard copy output. Printed output may be produced on high speed printers in a computer room or on slower devices at remote locations.

One disturbing feature of computers employed in commercial data processing installations is their tendency, despite the increasing use of VDUs for on-line information retrieval and computer output on microfilm (*see* next section), to output very large quantities of paper.

Reports, statements, invoices, purchase orders, production plans, analyses, and so on, appear from high speed printers at an alarming rate. The responsibility of the analyst does not end with the design of an acceptable printout. The analyst is also responsible for (1) ascertaining the numbers of copies required; (2) distribution of copies; (3) a system to provide the information in the physical form required with sprocket holes trimmed, continuous sheets separated, and interleaved carbons removed.

Fortunately, there is a wide range of machinery available to the user to enable the efficient performance of the third group of tasks. A generic name for this group is paper-handling equipment, and it comprises the following main subdivisions: (1) decollators;

(2) bursters; (3) guillotines; (4) trimmers; (5) combined machines; (6) paper-folding machines; (7) mailing equipment; (8) copying equipment.

Decollators are machines which will remove the interleaved one-time carbon from multi-part continuous sets, and separate the individual parts. These parts are refolded into individual trays, so that after decollation the original multi-part set becomes a number of one-part sets still in continuous form. The machines normally trim off the sprocket holes on both sides of the form. This latter operation can be performed on a trimmer independently, if required. Variants of the standard decollator are items of equipment which will remove one part from a multi-part set while leaving the balance of the set intact. Another device, called a recollator, performs a related operation of recombining the multi-part set in its original sequence after the carbons have been removed.

Once the carbon has been extracted, it is necessary to separate the individual forms from their continuous web. A machine which does this operation is called a burster. It tears the forms apart at their perforations and stacks them ready for distribution. One interesting adaptation of the device is the burster which is fitted with a signature plate enabling the separated forms to be signed before stacking.

Guillotines are normally employed as an alternative to bursters, and, as their name implies, they cut the individual forms apart rather than tearing them. Consequently, unperforated stationery can be accommodated when necessary. Guillotines can be used for perforated continuous stationery, in which case two cutting knives are fitted and these remove from between adjacent forms a very narrow horizontal strip containing the perforations. This provides a neater finish for the final document, as the rough edges of perforations left by bursting are eliminated. A guillotine can also be fitted with additional knives which will slit a document vertically, which is a necessary feature when two-up printing has been employed.

Most of these machines described above are available as free-standing units to perform the operations required to convert the printer output into acceptable documents for distribution. Many of the individual units can be linked together, however, in which case they are more properly described as combined machines. This arrangement allows the user to build up the paper-handling equipment to the exact specification required for a particular installation. Nearly all the machinery involved can be fitted with optional extras for jobs which are out of the ordinary, and the manufacturer's sales literature should be examined carefully before deciding upon the particular unit required.

When a neat stock of trimmed output has been obtained, the next problem is how to distribute it. Once more, specialized equipment is available to handle the task. Paper-folding machines are made which will fold individual forms to the required size to go into standard envelopes. More advanced models will collate and fold two or more forms together in the same way. To avoid the need for hand insertion into envelopes, some paper-folding equipment performs this job automatically. One feed hopper is loaded with the required output forms, a second hopper takes the envelope (in unmade form, normally) and the machine folds, inserts, seals and delivers the envelopes ready for posting. The only operation remaining is to stick on the stamps. If the volume of mail is large, this task can be made automatic by the use of a franking machine. A warning about taking care in the choice of envelope sizes is also relevant here. Future Post Office regulations may impose a surcharge on the use of non-standard envelopes, and this factor should be considered when discussing folding, inserting and other mailing aids.

At times, particularly with internal reports and documents, the problem facing the analyst is not one of distributing multi-output, but rather of producing multiple copies of one piece of output. Attention then centres upon copying equipment, and here the choice is considerable. There are two main ways of tackling the problem: (1) by producing the computer output on special 'stationery' which can be used as a master for subsequent copies; (2) by printing a single copy, on standard listing paper, and taking copies of this on conventional office copying equipment.

The use of special stationery covers output which is printed on translucent paper for dyeline copying, with hectographic carbon for spirit duplicating and on paper masters for offset duplicating. Each method has applications, and when justified on economic grounds can provide an efficient means of producing multiple copies.

If single copies are produced on the printer, the choice of copying processes is the same as is available for producing copies from any original office document.

One limitation on many office copiers, however, is the size of original document, since these are often designed for copying, as a maximum, A4 papers. Computer printout tends to be larger than normal office sizes and for this reason some documents are unsuitable for copying, although reducing copiers are available. Also manufactured are xerographic copiers which will accept computer output in continuous form, make reduced copies of individual pages, collate these and stack them ready for stapling or binding.

The final consideration of paper handling is how to store the printout when it is likely to be needed for subsequent reference. Specially constructed binders provide one solution; these have elastic cords which are threaded through the sprocket holes in the print margins and clamped under metal clips. This method of fixing allows the pages to be opened flat, so that none of the print is obscured in the paper folds. Other proprietary filing methods exist, and numbers of these are marketed by the continuous stationery manufacturers themselves.

Modern line printers can produce output at speeds of up to 4,000 lines per minute whilst laser printers can operate at speeds up to 20,000 lines per minute. It should be remembered that remote printing devices can have their speed limited by the transmission speed of the telecommunications link.

12.3 Storage devices

The vast majority of storage devices utilise magnetic media in some form, be it disk, diskette or tape. Semiconductor storage devices utilising silicon chip memory are also in common use but these have the disadvantage of being volatile, so that when the power is switched off, the contents of the memory are lost.

Hard disks are used for bulk storage of current data which needs to be accessed with little delay. Most of these devices are fixed in a sealed pollution free environment, together with the movable reading/writing mechanisms which access them. Such disk units may contain over 5,000 megabytes of data which can be transferred to and from main storage at some three megabytes per second. The storage capacities of disks attached to a minicomputer are usually measured in hundreds of megabytes whilst a microcomputer would typically contain between 20 and 40 megabytes of hard disk storage.

The way that data is stored on a disk will have a significant impact on the efficiency with which it can be used. If the disk contains many data files the space occupied by a single file may be fragmented across several different locations on different disks within the disk volume. This often results in considerable delay when the file is processed because of the need to frequently reposition the read/write mechanism.

Although most data files stored on disk are sequential in order, one of their main benefits is rapid access to random records. Thus, they are ideally suited to housing large on-line databases. Where both random and sequential processing of the data is required, the indexed sequential method is typically used. This means that the data is stored in the appropriate physical and logical order and the computer creates indices to enable the physical location of an individual record to be speedily determined.

A hard disk in a microcomputer will typically consist of just one rotating disk, rather than the disk volumes used with larger computers. Diskettes have the benefit of being inexpensive and portable but their capacity is usually limited to just over one megabyte.

Magnetic tape is also inexpensive and portable. A reel of tape costing about £12 can accommodate some 150 megabytes of data, depending upon how the data is organised. Tapes are used for sequential files and for housing backup copies of disk files. Data can be transferred between tapes and main storage at similar speeds to disk transfers, but searching a tape for individual records is a lengthy process. Most computer installations will keep an up-to-date copy of their disk files on tape within the computer installation, whilst a further tape copy of the most important files will be stored at a remote site for added security.

Many microcomputer users use a tape streamer to take backup copies of their hard disk on to a tape cassette. Copies can also be made on to diskettes but this is a time consuming process and involves much diskette changing. Equally, many minicomputer users forget to make adequate back up copies.

12.4 Data communications

There are many different ways to interconnect terminals and computers. Most data communications work involves linking a number of input/output terminals to a central computer of some description, but it is becoming increasingly popular for geographically remote computers to be linked together. The communications hardware providing these links however is similar. The specific hardware will depend upon a number of factors such as:

(1) The volume of data to be transmitted
(2) The transmission speed required
(3) The compatibility or otherwise of the devices at either end
(4) The degree of error checking required at the receiving end

(5) The confidentiality of the data being transmitted

(6) The degree of reliability and flexibility required of each device and each communications line

A typical communications link is shown in Figure 12.4.1, which shows a number of remote VDU input/output terminals attached to a single control unit. This control unit enables multiple terminals to share one communications line. The modem (an abbreviation of *mo*dulator/*dem*odulator) links the digital devices to an analogue telephone line. It translates the outgoing signals from a digital form to an analogue form and performs the reverse operation on incoming signals; thus a modem is required at each end.

The front end processor (FEP) looks after the data communications network on behalf of the host computer. This frees the computer to concentrate more on the processing of applications programs. The FEP typically performs the following tasks:

(1) Routing output to the appropriate remote device

(2) Accepting input from terminals and checking its validity before passing it on to the computer

(3) Performing code translation if the terminal and the computer use different data representation codes

(4) Inserting control characters around output data for addressing and error checking purposes

(5) Stripping these control characters from input data before passing it to the computer.

Another possible communications link is shown in Figure 12.4.2.

Note that in Example 2 there are a number of different types of terminal at the remote location which all need access to the host computer. They can all share one communications link through a multiplexor which provides the advantage of handling different types of terminal. However, a multiplexor will also be required at the computer end to split the signals into separate communications channels before they can be forwarded to the computer.

Another difference in Example 2 is that one of British Telecom's newer digital communications links is used. This means that unlike a conventional telephone line, the digital signals generated by computers and terminals do not have to be connected to an analogue form. This means that instead of a modem at each end, only a simple inexpensive interface device is needed. This device is supplied by BT with the Kilostream line and is called a Network Termination Unit.

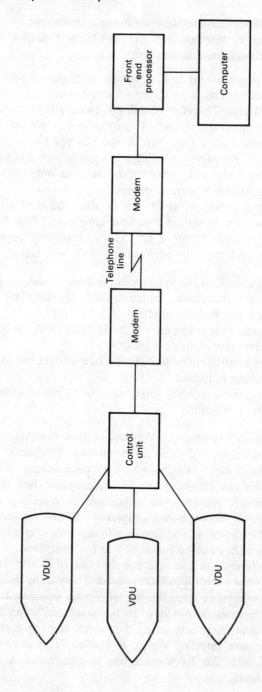

Fig. 12.4.1 Typical communications link (Example 1)

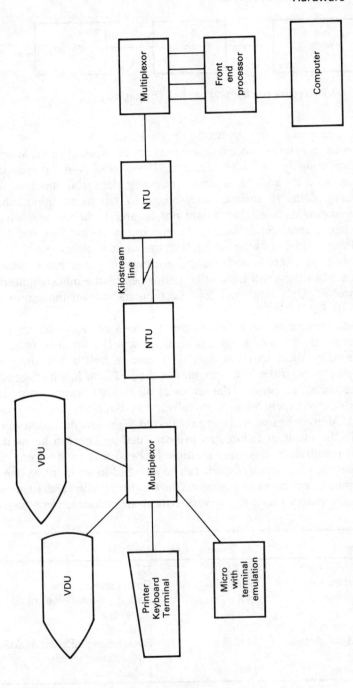

Fig. 12.4.2 Typical communications link (Example 2)

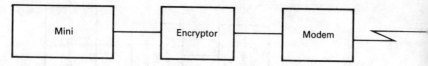

Fig. 12.4.3 Typical communications link (Example 3)

A further example is shown in Figure 12.4.3.

Example 3 shows two minicomputers linked together to exchange data of a highly confidential nature. Thus, there is an encryption device at each end to scramble outgoing data and unscramble incoming data. If anyone did manage to tap in to the public communications line, they would not be able to understand what was being sent. Sometimes data encryption is handled not by hardware but by software within the computers at either end.

Notice that there is no front end processor at either end because the FEP functions will have to be performed by the minicomputers themselves. Only large mainframe computers warrant the use of a separate FEP.

Each component of a data communications network will inconvenience one or more users if it breaks down. If a terminal fails, it will usually affect only one user. If a modem fails it will affect a group of users. If the host computer or the FEP fails it will affect all of the users. To provide duplicates of all the hardware devices is prohibitively expensive and unrealistic but flexibility can be provided. Many terminal users are connected via a private leased line but in the event of failure, an ordinary dial-up line can be used. Large mainframes may have multiple FEPs so that in the event of one failing, the connected users can be switched to an alternate one. Computers are becoming progressively more reliable and failures are increasingly rare but for very critical applications, some com-

Service	Maximum speed	Used for
Datel (dial-up)	9.6 Kbit/s	Data
Leased line	19.2 Kbit/s	Data, graphics
Kilostream	64 Kbit/s	Data, high-resolution colour graphics
Megastream	2 Mbit/s	Data, voice
Packet switchstream	48 Kbit/s	Data, linking different devices at differing speeds
Satstream	2 Mbit/s	Data, voice

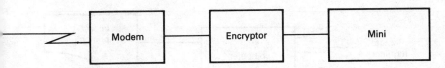

panies utilise fault tolerant computers which have all of their components duplicated. Some of these systems are claimed to have a total failure on average only once in every five years.

Although the analogue leased and dial-up lines are still very widely used, there are a number of newer digital services. The major British Telecom services are shown on p. 216.

12.5 Typical configurations

The system in Fig. 12.5.1 may have hundreds of local and remote terminals attached to it, plus perhaps fifty disk units, eight tape units and four high speed printers.

The system in Fig. 12.5.2 may have some fifty terminals attached plus perhaps four disk units and a single tape unit and high speed printer.

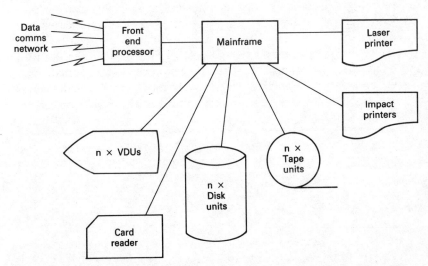

Fig. 12.5.1 Typical communications network configuration (Example 1)

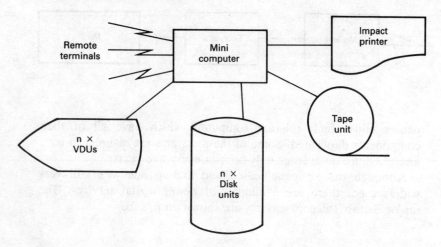

Fig. 12.5.2 Typical communications network configuration (Example 2)

13 Software

There are in existence many sets of coded instructions or programs which are designed to fulfil a wide range of commonly encountered requirements and make the best possible use of the computer hardware. Such sets of instructions are described by the all embracing name of *software*.

Much software is supplied by the manufacturers but, equally, a great deal of it is developed by the user to meet specific requirements or even purchased by the user from an independent outside body.

Regardless of origin, it performs two main functions. First, software is designed which enables the internal control functions of the hardware to be carried out efficiently at the speed of the fastest piece of equipment—the central processor. The second important function is to make the task of programming easier and this, in turn, reduces the delays before a user's machine can become fully operational and productive. Because software covers such a wide range of requirements, it is advisable to subdivide it into a few main types. The sections which follow discuss each of these in turn. It is necessary to realize that there is a considerable lack of standardization in software and some of it is peculiar to one machine only. However, certain general principles have been followed in its development, and this chapter is concerned with these: in this way it may be regarded as a survey of the subject, rather than a detailed specification.

13.1 Operating systems

As described in the previous chapter, the on-line computer hardware can be seen as a complex arrangement of individual units working together to manipulate data at very high speeds. To control the necessary actions efficiently, one group of software is concerned with the hardware organization. This group of software controls the input and output devices, internal data transfers and external communications, and ensures that required peripheral equipment is available for use at the correct times.

This facility becomes of vital importance when the multiprogramming concept is being used in the configuration as this mode of operation means that several running programs are in the computer at the same time. For example, one program may be sorting data, a second performing calculations and a third may be printing results of an earlier program from magnetic disc. All these programs are using the same central processor, but time on it is shared amongst all three according to successive relative priorities as they perform their allotted tasks. The whole sequence of decision-making involved in working a combination of programs successfully is controlled by special software, often termed an executive or supervisor program. The basis of the arrangement for sharing the processor time is that one program is allowed to run until either a natural delay occurs in that program, for example, initiating a data transfer to a peripheral, or an important event occurs that the processor should deal with immediately, for example, a hardware malfunction. When this happens, all the programs are examined and that with the highest priority is performed next until it reaches an interrupt point or until a previously interrupted job of a higher priority issues a fresh interrupt. The cycle is repeated until all programs have been completed. For obvious reasons, this method of control is usually called an interrupt system. Its use ensures maximum efficient use of the central processor.

A similar principle to that employed in the multiprogramming concept is used with interactive computer systems. In such a system, there may be many individual users connected to a central computer via keyboard terminals and all running different programs on the computer at the same time. It would be clearly unfair to allocate processor time on a priority basis since it is not known in advance what facilities and peripherals each user will require. Thus a time-sharing concept is employed. In such a system the processor time is available for a short time interval for each user before being terminated. This can be irrespective of the logical point that has occurred within the program's execution, but, in an interactive system, is frequently caused by data communication with the user. Once all the users have been serviced in this manner, the cycle starts again. Obviously in such a system, the more users there are connected to the computer, the worse the response time is for each user.

Another class of software, which falls within the overall definition of operating systems, is devoted to error detection, and this can take two forms. Errors caused by machine malfunctions will be found by this software. For example, if a parity failure occurs when items of data are read following transfer from one storage location to another, the software will cause the operation to be repeated in an attempt to clear the error. If it still remains, the program will be interrupted and a suitable error signal transmitted to the operator. Some software of this type has facilities for automatically printing out the entire contents of the internal storage, in machine code, when a failure occurs, so that the programmer can investigate the cause. This operation is commonly referred to as dumping, and, once again, is controlled by the software available. The main examples of interrupts are as follows:

(1) *Input/output interrupt* This denotes that a particular data transfer has been completed or that some error condition exists on the device or that an error has been detected as a result of the transfer.

(2) *Supervisor call interrupt* This is initiated by an application program which wishes to make use of a service provided by the supervisor routines or an input/output operation.

(3) *Program interrupts* These are caused by program abnormalities such as illegal statements or when data overflows the storage area allocated for it.

(4) *External interrupts* These are caused by remote or operator controlled devices (such as the console) which have been given an over-riding priority.

(5) *Machine interrupt* This is caused by equipment malfunction, such as parity failure or power loss to a peripheral device.

Dumping is also of considerable assistance to the programmer in detecting errors in the logic of his program. When faulty logic gives rise to recognizable error conditions within the machine, the software takes over and reports on the situation. Examples of this type of error are those caused by (1) registers being overfilled as a result of a calculation step; (2) incorrect modification of storage locations resulting in an attempt to transfer data outside the limits of a previously defined area; (3) attempting to use peripherals which are not attached. This group of software is closely allied to that designed for tracing the operation of a program. The diagnostic routines will print out selectively the results of each instruction as it is obeyed, the names of labels indicating the route being taken through a program, or the value of the contents of particular storage addresses, each together with a suitable commentary that the programmer can understand. With interactive systems, the user frequently has control of these diagnostic routines directly through his terminal keyboard.

Yet another part of the operating system software is devoted to communication with the operator. Most commonly such man-machine relationship is achieved via a console VDU, which displays a log commentary of significant event plus requests for commands. This log will include details of the programs being run, the peripherals in use, file identifiers and running times involved. When an error occurs, or a programmed interrupt point is reached, a suitable message will be displayed. Using stylized forms of address, the operator instructs the machine of the next action required by keying-in a combination of letters and numerals on the keyboard of the console. In some advanced applications, these enquiries and answers appear as normal English sentences, thus easing the task of the operator still further.

Operational efficiency is of great importance, since considerable time can be wasted in changing from one job to another. Within the operating systems, the type of software devoted to job control is aimed at minimizing this wastage. This type allows the operator to prepare future jobs while running current jobs. It will handle queues of programs, ensure that all necessary devices are available and attend to various housekeeping duties, such as controlling blocking and unblocking records on magnetic storage. This aspect of file control, which is part of job control, was discussed in Chapter 5. Another task is to ensure that the correct files are available for a particular job, and to ensure the operator has not loaded an incorrect one by mistake. In an organization with regular scheduled work, job control software can be instructed each morning as to what jobs are to be done, or even just what day it is, and it will control the whole sequence.

Two other terms associated with operating systems probably require special clarification: virtual machine and virtual storage. The virtual machine concept is the creation of an operating system which makes the computer seem other than it physically is to the user. Such an operating system can be employed to enable a user to access more main storage than is actually available (by using techniques described below), or even to use peripherals which are not connected to the computer by simulating their presence. Equally, a virtual machine operating system can enable many users to utilize the same processor in different processing modes, thus giving the appearance of a series of operating systems within a global operating system.

The virtual storage system is an operating system which allows a program to be executed on a computer which has insufficient internal storage available for that program. Equally, a virtual storage multi-programming system makes provision for running several programs at the same time, even though the total storage requirements for all of the programs may far exceed the amount of internal storage available. This is achieved by holding the programs on backing store, typically discs, and transferring small program segments called pages into main store for execution. The position and number of pages held in main store for any one program is dynamically adjusted to maximize the efficiency of the use of the processor.

We can summarize this section on operating systems by listing the minimum requirements and the additional desirable features.

Minimum requirements
(1) Load program into main storage.
(2) Allocate main storage and peripherals.
(3) Maintain continual supervision of program, including error dump and restart procedures.
(4) Maintain a log of important program events.
(5) Increase throughput by the elimination of operator waiting time yet also giving the facility for manual intervention.

Desirable features
(1) Better input/output control
(2) Automatic recovery from error situations
(3) Automatic scheduling on a priority basis
(4) Automatic accounting of time and resources used by individual jobs
(5) Multi-access storage protection
(6) Dynamic storage allocation
(7) Control of data transmission
(8) Control of a data-base management system

13.2 Common tasks

It was realized very soon after computers became of commercial significance that there were many tasks which had to be performed by every user irrespective of his particular business. Because it was wasteful for programmers in each installation to duplicate their efforts by writing individual programs to accomplish these tasks, the manufacturers supplied software with the machine to perform them. In the same way as the software for operating systems can be subdivided, so too can this broad category. Common task software can be considered as either organizational, system or programming.

Organizational common tasks are those which the user can select as being of use to his particular business. In this category are found the standard routines dealing with PAYE, Graduated Pensions, National Insurance and so on. Another common routine is the one which will perform a coinage analysis for a payroll. The distinction between these routines and those which will be considered later as 'application packages' in Section 13.4 is that the latter are complete in themselves. A routine for calculating PAYE, on the other hand, is designed to be incorporated as part of a larger payroll program. The software in this case simply saves the user programmer the labour of designing and coding what is essentially a standard calculation.

Common tasks that are associated with the system, and have their own software supplied, are sometimes referred to as utilities. In this group of software are programs for sorting and merging files into a prescribed sequence, and all of them are parameter driven because of their general nature. Also in this group, the routines which will edit data are found. Examples of this type are the insertion of currency signs, spaces, punctuation marks and other symbols into data prior to output, in order to make the printout legible. Other examples of editing software are the utilities in interactive systems for editing source programs from a terminal keyboard. Many other types of utility programs exist such as inter-peripheral transfer routines and report generators which enable the programmer to produce intelligible reports from data files with the minimum of effort.

In the programming area, where the common tasks software involved are usually called subroutines, sets of coded instructions are supplied which the programmer can insert into his own program. In this respect, they are closely linked with the organizational common tasks software already described. However, the organizational software is normally associated with a particular application (e.g. payroll), whereas programming software covers those tasks which are common to all applications. Examples of this types are the subroutines which calculate square roots, perform iterations or add days to the date. Many of the available statistical subroutines (e.g. calculating standard deviation) come into this category as well.

13.3 Programming aids

Another major group of software is designed to assist in the actual task of programming a system, which may or may not include some of the common tasks described above. Once again, a subdivision is possible between trial systems and languages. The former group includes all the specialist software which is available to the programmer to enable him to discover logical errors in his coding. It has been said that the only thing all programs have in common is the certainty that they will not work first time. For this reason the programmer spends a fair proportion of his time carrying out trials on small sections of the overall program to correct inconsistencies. When this job is finished, he must link the tested segments together to make an operational program. Finally, he must prepare test data carefully which will ensure that all the possible paths through the program are proven. At each of these stages, the trial system's software helps him and usually provides suitably documented reports of the actions that have been performed and the errors that have been found.

The other subgroup of programming aids—languages—incorporates a number of levels within it. The subject of software languages is very diverse, and many hundreds of different types have been evolved. There are, however, three main categories into one or other of which all languages can be classified. These three are machine, low level, and high level languages and in this order appear in increasing simplicity. Their purpose is common: to enable the user to communicate with the machine.

The computer hardware is activated by a series of impulses in binary form. It would be extremely difficult and error-prone to program a task as a set of binary coded instructions, and attempts to do this were discarded almost as soon as computers were introduced. When it was done the resulting program was said to be written in machine language, because the program bore a very close resemblance to the instruction patterns being obeyed by the hardware. It is very rare today for a program to be written in machine language, and it is only done when a programmer has a special task to perform which cannot be accomplished by other means.

Following closely behind machine languages came a large number of low level languages; almost every new machine appears with its own language. With a low level language, each function that the machine is capable of performing is given a mnemonic name that the programmer uses when he requires that particular function carried out. For example, he might write MOVNM when he requires 'move the negative value to memory', or SKIPLE for 'skip if memory is less than or equal to a specified value'. In each case these low level, or assembler, instructions would be followed by data addresses.

The computer accepts these mnemonics and translates them into appropriate machine codes. The use of these low level languages obviously assists the programmer, but he is still required to develop the logic at the machine level and consider the steps in operational sequence. The use of the language helps primarily in the coding operation.

This restriction brought the development of various high level languages which were problem oriented rather than machine oriented. With these, the underlying principle is that the user understands the problem he wishes to solve via the computer and can write this in a form which closely resembles a statement of the problem in English. There are certain rules which must be learnt, of course, before the high level language can be used, but in most cases these can be taught very quickly.

The main high level languages in common use are as follows:

(1) *COBOL* This was designed for commercial applications and is widely used on micros, minis and mainframes for business applications. It is an efficient language for file processing and input/output operations but is often criticized for being cumbersome and relatively difficult to learn. However, a survey in 1982 indicated that 70 per cent of programs in mainframe computer installations were written in COBOL. The language is available in many different versions, of which ANS COBOL is the most widely used.

(2) *FORTRAN* This was designed primarily for the easy expression of mathematical formulae and has been widely used for scientific applications. Several versions exist, of which FORTRAN 77 is the most widely used.

(3) *BASIC* Designed as an easy to learn and easy to use language which would enable the newcomer to use a computer with the minimum of learning. It has frequently been the only high level

language available on micros and has consequently been used for a wide variety of applications. However, it is also used on minis and even mainframes.

(4) *Pascal*   This is a general purpose language which can be used for mathematical and file-processing applications. It is particularly well suited for a structured approach to problem solving.

It should be remembered that owing to basic differences in the design of computers it is not always possible to incorporate *all* the facilities of a language into the program which translates the high level language into machine code. For this reason, it is not safe to assume that a program written in a high level language for one computer can necessarily be run on another machine even if that particular language software is offered on the second machine.

Despite the numerous problem-oriented languages which are in existence, there remains the problem of the programmer writing a program which is easy to understand and hence maintain or modify. It should not be forgotten that much of a programming team's time is spent engaged in program maintenance and so this task should be planned for at the initial program-writing stage. Initially good program documentation in the form of program flowcharts and the like were thought to be the answer, but now structured programming techniques are encouraged. The main objective of applying structured programming to data processing applications is to base the program design on the data structures used, utilizing the constructs of an independent high level language. Unfortunately, the more commonly used languages are not suited to forming these constructs, but this fact has not prevented the technique of structured programming being implemented commercially. It has meant that the final programs have become more readable and easy to understand and modify; hence more reliable.

13.4 Application packages

There are four main types of application package: system, commercial, engineering/scientific and mathematical. Systems analysts are concerned mainly with commercial packages, but may be involved in the acquisition of certain system packages, for example database management packages, decision table processors or transaction processing packages.

An application package is defined as being a complete system for a particular application and is supplied by some outside body for general use by individual firms. They use customer data to provide user departments with information and may be used by more than one installation or organization.

The benefits computer users derive from the use of application packages vary, but in general they fall into the following categories: (1) system implementation should be quicker and cheaper; (2) system efficiency should be better, based on the assumption that package designers will be more skilled than the average analyst; (3) good documentation is provided (although it is unlikely to be any one installation's standards); (4) easy to use on other machines in the same family or on a subsequent generation of machines from the same manufacturer (this is only likely to occur when using computer manufacturer supplied packages); (5) modification due to changes in taxation laws, introduction of metrication, etc., should be easy to implement.

Despite these advantages, many users are reluctant to use packages because of their alleged inflexibility and apparent difficulty of implementation. A thorough evaluation of the costs of implementing a package and of its performance, as against implementing one's own system, can often show that an application package is a 'better buy'. As package design continues to improve and installation labour costs rise, this has been and will increasingly continue to be the case. There are limitations, however, and these have to be borne in mind.

Sometimes the user's clerical systems require modification to tailor them to the requirements of the package. Other limitations become clearer if we examine the characteristics of a modern package:

(1) *Modular design with structured coding* The idea is to provide program segments that are reasonably well self-contained so as to provide the user with the opportunity to build in his own options and to faciliate program modification. There is a danger, however, that modularity could imply duplication of programming between modules. Structured coding is becoming increasingly important to aid the readability and understanding of the software.

(2) *System changes* Three situations may call for change: initial changes, changes due to program maintenance and changes due to expansion of the package's facilities. It is important to establish who is responsible for these changes, and should it be the user then source listings should be insisted on.

(3) *Documentation*   The quality of documentation provided with application packages varies considerably. Anything less than full documentation should not be accepted. Although in practice it is unlikely that this will be the case, ideally the following seven areas should be fully documented: overall system flowchart; identification of input and output elements; file record specifications; program logic; data specifications for each record type specified in the programs; test data; recommended clerical procedures, operating procedures, and conversion procedures.

(4) *On-site assistance*   The supplier should specify the support which he will provide with his package. It should certainly include help during conversion and initial running together with some formal education programmes for the customer's staff.

(5) *Payment and running costs*   The aquisition costs and methods of financing vary tremendously across the wide range of packages that are available. Some packages are offered only for purchase by the user, others for leasing, and some can be acquired by either method. If maintenance support is available with the package, it is usually inclusive in the leasing contract, but is an additional cost when purchasing and must be regarded as an extra running cost.

The implementation of an application package is a major exercise and should be treated as seriously as would the implementation of an installation-designed system.

Although the installation will be using fully written and tested packages it will still need to run a number of tests for its own benefit. These will include (1) test running, which will be controlled by the programmer, with data provided by the analyst; (2) pilot running, controlled by the analyst and based on data provided by the user department; (3) parallel running, which will be controlled by the user on live data used by the system which the package will replace.

Examples of applications which have been covered by packages include the following: modelling and simulation; computer-aided design; production control; stock control; payroll; project management; order processing; accounting; costing and pricing; statistics; numerical control; ledger accounting; personnel.

The quality of packages has generally improved and is expected to continue to do so in what has become a very competitive marketplace. A trend is developing for the introduction of integrated packages. A user organization can thus acquire selected packages and, as he expands his computer configuration, he can build up a fully integrated data processing system. Indeed many small business machines and microcomputer systems are sold as complete systems with application software. This is discussed further in the following chapter.

13.5 Fourth generation languages

Programming languages have changed over the years as more emphasis has been placed upon greater programmer productivity. Languages have become less machine orientated and more people orientated.

First generation programming languages were pure machine code and required an intimate knowledge of the computer in use. Second generation languages are typified by Assembler and, although close to machine code, they do permit the use of English-like mnemonic instructions. Third generation languages are the most widely used — the best examples being COBOL, FORTRAN, PL/1, BASIC and RPG. However, third generation languages were designed for programmers who would draw the flowcharts and then write their code at their desk. Fourth generation languages were devised to be easier to learn, more automated and to make applications programmers far more productive.

A fourth generation language (4GL) is much more than just a language, it typically consists of:

(1) A non-procedural language
(2) A report writer
(3) A query language
(4) A screen mapping facility
(5) A decision support language, and possibly
(6) A database management system

A non-procedural language is a coding system or command structure that enables the user to specify requests and instructions in a free form English-like manner. Thus, the programmer has less worry about writing each instruction to a precise set of syntax rules and many of the instructions are automatically generated by the 4GL software. Thus, the programmer devotes more thought to the problem at hand and less to the coding. The report writer allows the user, who may not be a programmer, to format an output report with the appropriate headings, totals and spacings, etc. The query language is designed for easy extraction of data from files and/or databases. Again, this may be used by programmers and end users. A screen mapping facility enables the programmer quickly to define screen layouts, perfect them and then file them — all without the need to code any instructions. The decision support language enables an end user to store and retrieve data and to organise data into a mathematical or statistical model for analysis.

4GL products are now used for many new application development projects as a means of involving the end users more. Vendors of 4GL products claim productivity improvement factors of between two and ten. They also claim that fewer high-skilled coders are required, the resultant systems are easier to maintain and permit end-user involvement all the way through the development process.

However, any computer installation that acquires a 4GL simply as a means to produce COBOL-type programs more quickly, without adopting the other 4GL facilities, is only likely to produce lots of bad software very speedily. The use of 4GLs does not compensate for poor system design. It has been estimated that for every £1 spent on new system development, a further £9 will be spent on maintenance during the lifecycle of that system. What the computer industry calls maintenance is really 'correction', often of gross errors made during systems analysis.

IBM, for example, estimate that if it costs £100 to fix a problem during the analysis stage, it will cost £10,000 to rectify it after the implementation of that system.

Using fourth generation techniques, analysis should begin by producing, amongst other information, two fundamental business models:

(1) The entity model showing the important things within the business about which data may be recorded.

(2) The function hierarchy showing a list of the basic business functions which are needed to meet the business objectives.

These models are absolutely fundamental to the fourth generation of computing. Any business function, for instance invoicing or stock recording, can be defined in terms of its relationship to the entity model, i.e. what it does to one or more entities.

Put together, the entity model and the function hierarchy illustrate the way that the data and information resources of the business are organised and manipulated to support the business objectives. This type of modelling process is the key to fourth generation systems development. Effective user involvement is the primary means of effecting quality control over the modelling process and to help refine the models into operational computer systems.

These tools are designed to automate the systems analysis and design and, in theory, once the systems design work is completed it will only be a matter of 'pressing the button' to automatically generate all of the live programs needed to make the system function. Needless to say, life is more complicated than this.

14 Micros

All areas of information technology have been affected by developments in microelectronics over the last few years — what is sometimes termed the microprocessor revolution. Not only has the advent of the microprocessor made more processing power available for less cost, it has provided cheap, simple solutions to the end user.

This chapter describes how the development of microcomputers has changed the range of solutions available to systems analysts.

14.1 Microelectronics and microprocessors

Early (first generation) computers were based on thermionic valves and were big, consumed a great deal of power and were often unreliable. Even so, the computing power they made available was revolutionary, mainly in science, but also in some commercial applications. The invention of the transistor led to the second generation of computers, consisting of separate transistors, resistors and capacitors. Compared with the first generation, these computers had greatly increased reliability and power, and were less expensive and smaller. Nonetheless, a computer based on this second generation technology was still a very expensive item and had to be regarded as a corporate resource, shared by many people.

A very significant advance in electronics was to fabricate a number of components in a single wafer — chip — of a semiconductor, usually silicon, and this led to the third generation of computers which utilised such integrated circuits. Integrated circuits are now available for many electronic applications, including amplifiers, computer memory and processors. Typically, each chip is a few millimetres long, but is often not visible, as it is enclosed in a black plastic package with metal 'legs' for insertion into a circuit board.

The number of circuits that can be put on a single chip has increased rapidly, and the various densities are referred to as small scale integration (SSI), large scale integration (LSI) and very large scale integration (VLSI). This corresponds to about 10, 1,000 or 100,000 circuits on a single chip. Currently (in 1987) memory chips are available with 1 Mbit capacity. Hand in hand with the reduction in size, integrated circuits have become faster, so that circuit operation may take only a few nanoseconds, allowing microprocessors to deal with more than one, and sometimes very many, instructions in each microsecond.

One very important type of chip is called a microprocessor. Compared to a VLSI chip, it has a simplified architecture and is shown in Figure 14.1.1.

Early microprocessors—for example the Intel 4004—are referred to as 4 bit microprocessors because they have registers which each hold 4 bits and a data highway which is 4 bits wide. Four bit microprocessors are still commonly used as controllers in products such as washing machines. Systems of this type are usually referred to as embedded systems and are not dealt with in this book.

The development of 8 bit microprocessors such as the Intel 8080 (1974) and Zilog Z80 (1976) heralded a new era by making it possible to build a general purpose computer around the microprocessor.

Other components for storing data and programs, input and output are, of course, required and a schematic logical structure for a microcomputer is shown in Figure 14.1.2.

The earliest such computers were built by electronics enthusiasts, but standard computers were soon available as complete products and many new companies sprang up to produce machines for this rapidly expanding market.

The next generation of microprocessors, with 16 bit registers and a wider address bus allowing more memory to be accessed, started becoming available in about 1979. Some of these microprocessors had an 8 bit address bus—for example the Intel 8088—and some a 16 bit data highway—a true 16 bit microprocessor. The Intel 89088 might be referred to as an 8/16 bit microprocessor and the others, for example the Intel 80286, as a 16/16 microprocessor. At the present time (1987) each of these is available in several formats up to 32 bit registers and data highways.

14.2 Typical business microcomputers

The commonest type of business computer is an IBM PC or compatible. 'Compatibles' are manufactured by companies other

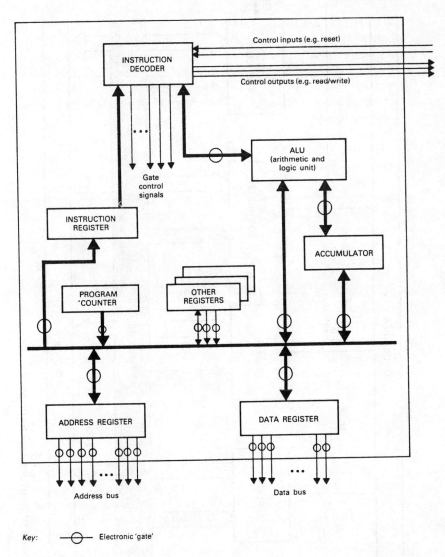

Fig. 14.1.1 Simplified microprocessor architecture

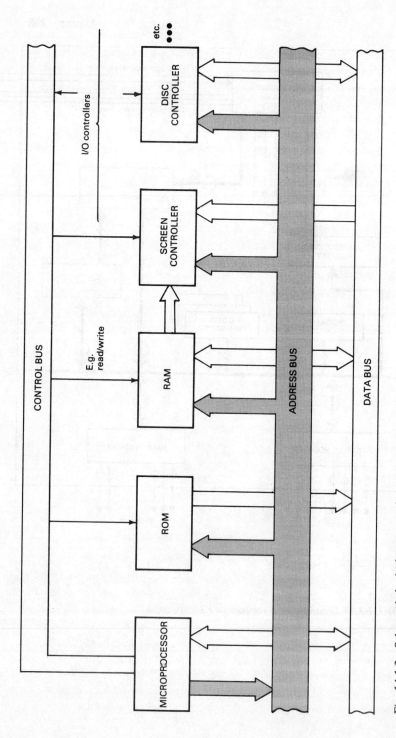

Fig. 14.1.2 Schematic logical structure for a microcomputer

than IBM, but provide a similar hardware and software environment. Many companies have sprung up in this area of business and some have been able to compete directly with IBM for performance and quality. Most compatibles sell on their price, considerably undercutting IBM.

Other non-compatible machines such as, for example, Apricot have failed to capture a foothold in the market and manufacturers have now generally converged on the IBM PC range as a de facto standard. The IBM PC standard is so universal that it is even possible to buy standard cards to provide upgrades to IBM-compatible PCs for more memory or better graphics. One recent innovation is the possibility of using some very fast microprocessors such as the Inmos transputer, or 'reduced instruction set' microprocessors on 'drop-in' cards within a standard PC. This allows very powerful computing with many millions of instructions per second to be made available for some special problems at the fraction of the cost of a mainframe solution.

A typical business microcomputer consists of:

(1) An industry standard microprocessor
(2) RAM—random access memory for holding data and packages
(3) ROM—read only memory for holding some operating system components
(4) Display screen—to display text and graphics
(5) Sound output—a simple output to indicate warnings is normal, but 'voice output' is also available
(6) Input devices—a keyboard is universally provided, but the use of other devices such as a wand or a mouse is increasing
(7) Storage devices—for storing large amounts of information for later retrieval
(8) Communications ports—to drive printers, link to other computers and allow access to remote devices via a modem

Desk-top microcomputers plug into a mains socket and consist of a keyboard, display unit and system box containing the microprocessor and memory, etc. The power supply unit converts local power voltages into a suitable internal voltage for the digital components of the computer. A 5 volt supply is normal, but 12 volts may also be required for communications ports.

Portable micros are similar, except that they are made to pack away into a single unit for transport. They usually include a small screen, but need to be connected into a mains unit before use.

The industry standard microprocessor is the Intel 8088 found in the IBM PC/XT or the Intel 80286 found in the IBM PC/AT. Other manufacturers have produced 'compatibles' using the same microprocessors and some are now using the Intel 80386. It is standard to provide at least 512 kbytes of RAM as many common packages require at least this much. It can be extended to several Mbytes for memory intensive applications.

Displays are one area where technology is changing rapidly. A few years ago microcomputers could only display text, but now they are able to display graphics with horizontal and vertical resolutions approaching 1,000 pixels, either in monochrome or colour. This has made microcomputers particularly useful for graphics workstations attached to larger computers, as well as in stand-alone applications such as simple statistical charting or electronic publishing. Most displays utilise cathode ray tubes as in televisions, but liquid crystal or plasma display flatscreens are increasingly seen, particularly in portable machines.

Window, icon, mouse and pull-down menu (WIMP) interfaces are now found on all microcomputer types. Available commands are represented by icons displayed on the screen. An icon is a miniature picture representing a system option and the mouse is pushed around a flat surface to indicate which icon is to be selected. For example, pointing at the name of a file and then at a picture of a waste-paper bin deletes that file. A window is a section of the screen dedicated to one function and several windows may be in use on the screen at one time. For example, when replying to an electronic message we might display the message in one window, a list of all files in another window, and the reply being created by a word-processor in a third. The individual displays may be scrolled within each window, so that we may reply to a long memo. New windows may be created, or 'pulled down' to overlie existing windows, but when we have finished with them they may be released, revealing the window that they were overlying in unchanged form.

Floppy disk drives are universally provided with business microcomputers so that data and programs can be permanently stored and transferred to other machines. The standard PC disk is 5.25 inches in diameter and can store about 0.5 Mbyte, but the use of 3.5 inch 'microdisks' is on the increase. One great advantage of floppy disks is that they are exchangeable and a large library of disks can be kept. When a larger volume of data needs to be stored or the continual changing of floppy disks is inconvenient, a hard disk may be added to the machine. Hard disks, which were formerly known as Winchesters, allow storage of 10 to 20 Mbytes. Hard disks allow considerably faster data storage and retrieval, but as they are not exchangeable, require to be regularly backed up onto some exchangeable medium and held off-line. This can be done by using a large number of floppy disks, but special digital tape units called tape streamer devices are specially designed for this purpose. Storage devices based on optical techniques as used in compact disk players are now becoming available. These allow storage of up to 500 Mbytes, but data is read only (CD-ROM) or may only be written once (Write Once Ready Many — WORM).

The ability to print letters and reports is provided by means of printing devices such as daisy-wheel, matrix and laser printers. Daisy-wheel printers provide letter quality output for text, with typical speeds of about 40 characters per second. Matrix printers are able to print text and graphics with a resolution of about 100 dots per inch. Matrix printers are faster than daisy-wheel printers, but do not offer such high-quality output. Both are noisy requiring acoustic shielding in an office environment and are increasingly being supplanted in the market-place by laser printers, which are much quieter and quicker, offering letter quality text and graphics with a resolution of 300 dots per inch. Where colour output is required, graph plotters may be used, but there are also devices for producing 35mm slides from the screen display providing much better quality than would be obtained from photographing the screen.

It is very important to realise that a microcomputer is just so much useless hardware unless a suitable operating system is used. This is also true of larger computers as well of course, but the development of industry standard operating systems that run on computers from different manufacturers has begun to lead purchasers of computers to demand independence from the vendors of proprietry software. An early example of this was CP/M, produced by Digital Research and capable of running on any Z80 or 8080 based machine with disks. MS-DOS from Microsoft and the equivalent PC-DOS were developed as the operating system for IBM-PCs and compatibles.

It is in the interest of the larger manufacturers to resist the establishment of industry standards, because they can then be undercut in price by other manufacturers offering the same products. On the other hand, however, purchasers prefer industry conformity, which avoids becoming locked in to a particular supplier and also facilitates planning and makes training easier. Software houses are also able to offer a wide range of high quality software that runs in such standard machine environments. In the long run, of course, the purchaser has the ultimate power and it is now taken for granted that industry standard operating systems will run on hardware from different manufacturers. In the mini and mainframe market, it is now possible to buy UNIX machines supporting large numbers of on-line terminals and running standard packages such as ORACLE and INGRES from several suppliers. Users know that this philosophy provides them with security of supply as well as guaranteeing a source of usable, debugged packages running in standard operating environments.

14.3 Typical applications of microcomputers

There are very few single-user computer applications that cannot be accomplished using a microcomputer. Typical of the wide range of uses that they are put to are the following:

(1) *Word-processing*: here the microprocessor is used to run a program such as Wordstar, Displaywrite, etc., that allows creation, editing and printing of text documents. In the early 1980s word processing was carried out on special machines but the use of industry-standard microcomputers running relatively cheap packages is now almost universal. These wordprocessing programs often include such features as spelling checkers and the ability to use files created by some other types of packages.

(2) *Database applications*: here the microcomputer acts as a traditional data processing computer. Data is stored in files and accessed by database software in the same manner as on larger machines. Users are able to update information and produce reports using the facilities provided by the database package without having to worry about the complexity of file handling. There are three types of database found on microcomputer systems. The first type regards data as being held as a series of 'flat' files, with facilities to edit and display or print the files.

A second type provides increasingly sophisticated implementations of a relational database, with extensive programming support to build display screens for standard enquiries. The most well-known product of this type is dBase III, costing a few hundred pounds in the UK. Update and enquiry functions may be produced by including database statements in a program written using a standard high-level language such as BASIC or Pascal. Sometimes a special database language may be provided. Non-programming techniques, such as screen painting to define the layout of reports, may also be provided.

The third type of database found on microcomputers is exemplified by ORACLE and INGRES, both of which are powerful relational database products that run on a variety of machines from mainframes down to PCs. Access to the database is by means of the same user interface, whether the machine being used is a large IBM or DEC machine or a relatively humble PC. This has obvious advantages for user training and application development. Future trends in corporate information handling and retrieval will include linking PCs into wide area networks with data being accessed from any node on the system. Users will not need to be aware of the geographical location at which data is being held or the machine supporting that part of the database.

(3) *Spreadsheets*: here the computer shows a rectangular table of data which may be manipulated or printed. Columns or rows may be added to produce marginal totals, such as monthly sales figures, by very simple commands. Financial projections may be carried out by, for example, multiplying a row of monthly sales figures by 1.12 to allow for annual growth. This type of very simple data processing facility is often a very cost-effective first contact with data processing for managers, but is capable of quite sophisiticated applications such as data importing from databases. There are many spreadsheet packages in the market place, two of the most widely known being Supercalc and Lotus 1–2–3.

(4) *Computer aided design*: here the microcomputer is used to set up and edit a graphics database. The higher resolution graphics now available make this increasingly cost-effective for a wide variety of end-user requirements. Standard engineering applications are an obvious area for CAD, although the microcomputer is usually linked to a larger machine, which calculates stresses and performance and may even generate output files to control machine tools. Computer packages are even available for such humble tasks as designing knitware!

(5) *Electronic publishing*: this combines text and graphics and allows the generation and production of technical manuals using relatively cheap facilities based on the microcomputer. A system of this type will include a microcomputer running a standard package, such as Ventura, which allows integrated text and graphic documents to be produced. In order to produce quality output a good quality printer is required and this is often a desk-top laser printer,

but the output from the better packages of this type may also be used to drive phototypesetting equipment to produce book quality output.

(6) *Dedicated packages for one application*: examples of this type of package include complete accountancy packages for a small business, including all aspects of accounting which used to be carried out manually. A systems analyst would be well advised to consider these when looking at small applications.

(7) *Specific programs*, written using standard high-level languages such as BASIC, Pascal or C. Compilers are available to support most languages but this is rather against the accepted philosophy of how microcomputers should be used within a business. Good software may consume resources alarmingly—for example, one week of a programmer's costs would pay for a word processor, a database package and a spreadsheet!

14.4 Local area networks

The first microcomputers were used in a stand-alone mode with a communications capability that was used only to drive printers attached to them. Enthusiasts, however, soon found ways of connecting micros to mainframes to emulate standard remote terminals. Primitive ways of transferring data between one microcomputer and another were devised using the serial communications port originally provided for the printer. The software to achieve terminal emulation or data transfer was fairly primitive and was designed to solve a specific problem, but solutions to these requirements are now available as standard products, both hardware and software.

The reasons for wanting to link microcomputers together and to link to mainframe computers fall into three general categories:

(1) *To share resources* Some resources such as printers may be relatively expensive compared to the cost of the microcomputer. A desk-top laser printer, for example, may cost more than the micro itself and it clearly makes sense to share this resource amongst several users. One solution would be to provide a separate machine that is only used for printing. When someone required a high-quality printout, the file could be transferred to that machine by means of an exchangeable floppy disk. There may be several people waiting to load their floppies into the printing machine at any one time, however, and this queuing for resource is wasted time as far as they are concerned. A networked solution allows files to be trans-

mitted to a shared printer so that any required queuing is carried out by system software. This allows the user to get on with other jobs while waiting for the file to be printed.

(2) *To communicate* One longstanding myth is that the microprocessor revolution would lead to each employee having their own microprocessor and all communication would be by means of electronic mail, thus leading to the paperless office. This has not materialised—in fact, because of the greater ease of document editing and production, and the advent of faster printers, more rather than less paper seems to be used by many organisations. Some companies have utilised electronic mail more than others, particularly multinational companies, where the avoidance of failed telephone calls because someone is unavailable or where time differences make direct contact difficult, means saved time and money. Messages are left in an electronic mailbox for the recipient to collect when they log onto the system.

(3) *To control how people access files and programs* If several people need to use the same information in an organisation, they could each be given a micro and a floppy disk containing the required files. If the information does not change, then this is a satisfactory solution, but as we all know, the only thing we can guarantee will always be the same is that things will change! In the case of programs or data files which only change relatively infrequently, a viable solution would be to issue a new version of the disk from a central control point. A package to calculate tax liability could be re-issued yearly and such commercial pachages are available. In a similar way, some companies offer a maintenance contract for information services with CD-ROMs being sent to subscribers on a regular basis.

For data that changes over a shorter time-scale, and this is the case with most database applications, the transporting of data via a physical medium is not appropriate and it must be centrally maintained and updated. Access to it is controlled by software running on one machine—the file server or database machine, depending on the context—but users may access it via terminals or other computers. It is possible for the microcomputer to download files from a file server for use in the micro and upload files to be saved in the opposite direction. These files could consist of data that is to be mainpulated by means of a spreadsheet, or it could be a centrally stored program that may run on any of the microcomputers on the network.

In the late 1970s, computer manufacturers such as Digital Equipment Corporation produced software that allowed several of their computers to be connected together into a Local Area Network (LAN), but problems arose where users wanted to connect together machines from many different manufacturers. For any one pair of machines, it was possible to solve the problem, but what was wanted was some agreed standard for interconnection. The protocols adopted by the larger manufacturers were one obvious possibility (Ethernet, originally designed by DEC and Xerox is an example of this type of standard).

The main thrust in this area, however, has been the work of the International Standards Organisation (ISO), the Institute for Electrical and Electronic Engineers (IEEE) and the International Telegraph and Telephone Consultative Committee (CCITT). ISO developed a theoretical reference model — Open Systems Interconnection (or OSI) — for connecting computers together. The OSI model does not define any physical standards, so that there is no intention that products meet it as a standard, rather it is a standard way of thinking about the problem. IEEE has produced a series of standards that are physically realisable in conformity with this model and actual products, both hardware and software, meeting IEEE standards are available in the marketplace. Two types of protocol covered by IEEE standards are Carrier Sense Multiple Access with Collision Detection (CSMA/CD) and those based on Control Tokens. Ethernet is an example of a CSMA/CD approach and many suppliers including IBM provide token passing ring network facilities for PCs.

In token passing networks, the network has a permit token which is owned by only one transmitter at a time. A device may only transmit onto the network when it is in possession of the token, in a similar way to the approach used to control access to single track railway lines. IBM offers a token ring approach of this type, using broadband coaxial cabling. Broadband cabling allows many communications channels to exist using the same cable, and these might be used for video, telephone conversations, etc., as well as data transmission. Baseband cabling, which is cheaper, only allows a single communications channel for the data.

A typical company might have several Local Area Networks (LANs), connected together as in Figure 14.4.1. LANs may also be connected to mainframes or remote systems via telecommunications links as shown. When machines are connected via telecommunications links, the term Wide Area Network (WAN) is used. This is

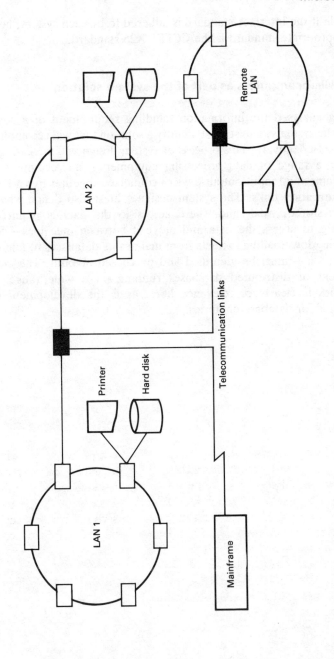

Fig. 14.4.1 Interconnection of LANs

possible if the interface standard is adhered to by each system, here the appropriate standard is the CCITT X25 standard.

14.5 Microcomputers as part of the system solution

Having analysed the information handling requirement of a company, the systems analyst must identify a solution that will accomplish this most effectively. This process of system design will increasingly involve a range of data processing equipment — microcomputers, mainframes and input-output devices connected together by various communication links. The system designer must also define where data is stored, used, and where access to the data is required. Software to access the data and move it between machines — for example, downloading extracts from mainframe databases to micros for analysis — must be identified and purchased or built. The development of distributed databases running on a wide range of machines is clearly of relevance here, as is the development of standards for database enquiries.

15 Introducing Structured Methods

So far in this book we have concentrated on techniques for analysis and design: we have discussed interviewing and decision tables, file design and timing. In this chapter we want you to step back from the detailed material covered so far and consider the overall methodology behind systems analysis and design. In particular, we want to consider structured methods, but before saying what they are let's take a pessimistic view of systems analysis and design to date.

Historically, systems development projects have resulted in over-budget systems which have not always met the user's needs, nor achieved the objectives set out for them. This should not be surprising. The development of new systems is a complex task and while very many new computer-based systems do work and are used, we know from our experience in developing them that it could have been done better. Additionally, the complexity and variety of systems to be developed is increasing and we suggested in Chapter 1 some of the new dimensions to systems work which will be apparent during the 1980s. It is painfully clear that traditional methods of carrying out systems analysis and design will have to change. Above everything, the systems analyst needs a formal, disciplined approach to analysis and design which will enable users to play their full part in the systems development process and ensure that new systems are developed in a cost-effective way. We believe that structured methods provide this approach. These structured methods include the use of:

(1) structured analysis
(2) structured design and
(3) structured programming

Each of these topics will be addressed in this chapter. Since structured methods were first developed in a programming context, let's begin there.

15.1 Structured programming

In the 1960s in the United States a number of surveys showed what most data processing managers had believed for a long time — that there is a substantial variation in programmer abilities and that too much time is spent on debugging programs and on maintenance activities. The full surveys generated much controversy, but the effect they had was dramatic. Suddenly everyone was concerned with programmer productivity and began to examine the way in which programmers programmed. In 1965, Professor Dijkstra of Eindhoven University in Holland presented a paper at the IFIP Congress in New York suggesting that the GOTO statement should be eliminated from programming languages altogether, since program quality was inversely proportional to the number of GOTO statements in a program. In the following year, Böhm and Jacopini showed that any program with single entry and exit points could be expressed in terms of three basic constructs:

(1) sequence or carrying out a process
(2) iteration or looping
(3) selection or decision taking

In familiar format these are shown in Figure 15.1.1.

From these beginnings was structured programming born. This dramatically improved the quality of programming and of programmer productivity.

There is no doubt that structured programming has been successful, but it doesn't solve all of our problems. Poorly constructed system designs can still negate the benefits provided by structured programming. Not surprisingly then, similar principles were applied to the tasks of analysis and design and a full range of structured methods came into being.

15.2 Structured analysis

To produce well-structured systems designs, the analyst needs to define accurately the outputs from the system, its inputs, the data structures and the processing. To help him do this, information is gathered from a variety of sources in several different ways. Further, the analyst checks back with many different people in different circumstances to ensure that what he is doing is right. The output from this process is a systems specification which the user is ex-

●Sequence

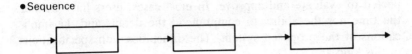

●Iteration

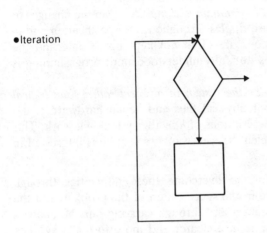

●Selection

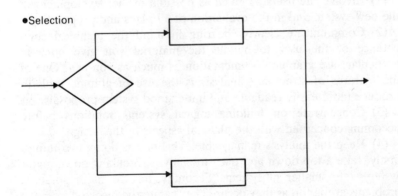

Fig. 15.1.1

pected to evaluate and approve. In most cases, users have neither the time nor the ability to comprehend the details and the implications of the proposed system. The difficulties with specifications of this kind are:

(1) *They are too big and complex.* It is quite unreasonable to expect users to search through long and complex systems proposals in order to verify the details of the systems proposed.

(2) *They are difficult to maintain and modify.* Simple changes to requirements often cause ripples throughout the specification and in consequence specifications are often not kept up to date and the implications of changes not fully understood until programming is reached.

(3) *They often describe the system in physical rather than logical terms.* This means that physical files and actual hardware is described together with descriptions of *how* the system will work. This muddles the picture which the user wants to see; he is interested in *what* the system will do.

Structured analysis aims to overcome these difficulties through involving the user more in the specification of the problem and the development of the solution and through carrying out the analysis and presenting the results in a clearer and more formal way. The following guidelines have been developed to illustrate this process.

(1) Involve the users as much as possible in the development of the new system and in the evaluation of all plans and proposals.

(2) Communicate clearly, bearing in mind the technical competence of the user to handle the material you give him. In particular, use graphic documentation as much as you can. One of the features of structured analysis is the use of graphic tools to produce more easily readable and understood system proposals.

(3) Concentrate on building logical systems solutions before becoming concerned with the physical aspects of the design.

(4) Keep the analysis manageable. This means doing two things: firstly, take a top down approach to analysis, breaking down major systems into smaller subsystems. Secondly, resolve major potential problems at the top as they occur. They cannot be resolved from the bottom up.

There are a variety of tools available to help us in structured analysis. These are:

(1) Dataflow diagrams. These provide a logical model of the system and show the data flow and the logical processes involved.

(2) Data structure diagrams which describe the logical data structures in the user's system.

(3) A data dictionary. This is the organized collection of all the data names used in a system and shown in dataflow diagrams or in data structure diagrams.

(4) Structured text which describes in a precise but readable way what actually happens in the lowest level dataflow diagrams.

Let's look at each of the above in more detail:

Dataflow diagrams (DFDs) The very simplest dataflow diagram might look like Figure 15.2.1; it describes the logical process of making a dry martini!

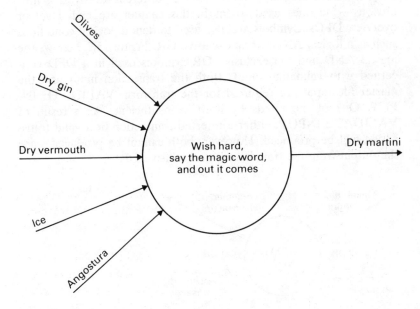

Fig. 15.2.1

While this DFD obeys some of the rules for dataflow diagrams in that it shows the inputs and outputs clearly, it's not very useful in the production of dry martinis to a consistent quality. It's rather like the one in Figure 15.2.2 for a customer enquiry system.

Dataflow diagrams are used to emphasize the logical flow of data through a system. The basic symbol is a circle — or bubble — and is called a 'transform' since it identifies a function that transforms

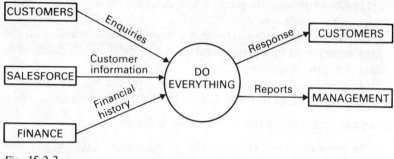

Fig. 15.2.2

data. Dataflow diagrams can be done at different levels, as is now obvious—the ones used so far in this chapter are high level or overview DFDs. Symbols are also used to denote logical conditions such as a logical AND, or an exclusive OR. Figure 15.2.3 shows the logical AND and the exclusive OR symbols used in a DFD concerned with validating input. Both the transaction input and the master file input are required for the transform VALIDATE INPUT. One of the inputs by itself is insufficient. As a result of VALIDATE INPUT, either a rejected transaction or a valid transaction will be produced. However, both cannot be produced since the output from the transform is an exclusive OR.

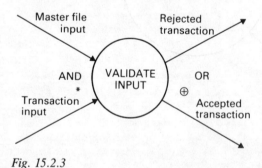

Fig. 15.2.3

The complete picture of the elements of a DFD is shown in Figure 15.2.4. In addition to the bubbles, we have SOURCE, DATA STORE and DESTINATION.

A simple customer ordering process might be represented as in Figure 15.2.5; it has the same component parts—source, process,

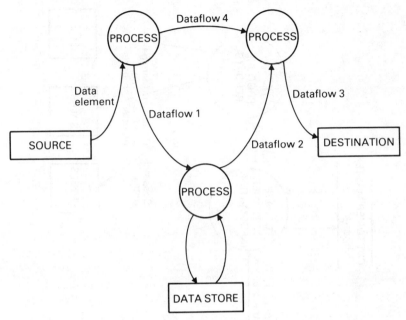

Fig. 15.2.4

data store and destination, but this time they are labelled in a form
appropriate to the application.

Data structure diagrams A data structure diagram describes the
user's logical data structures. To understand these logical data
structure diagrams (LDSDs) we need to become familiar with some
new terminology. Imagine a logical data structure for a local tennis
league which the league secretary accesses by team name and player
name. The data structure might look like Figure 15.2.6.

In this diagram, two data *entities* are shown: a TEAM and
PLAYER. Each *record* within the TEAM entity is uniquely identified by the name of the team, and TEAM-NAME would be a *key
attribute* since it would be used as a search argument to identify a
particular logical record within the entity. In other words, the
league secretary would use it to locate information about the team.
Each record in the PLAYER entity is identified by PLAYER-
NAME, the key attribute for that entity. By convention, attributes
which point to other entities come last, so PLAYER(S) is shown at
the end of the attribute listing for TEAM because it has a *logical
pointer* linking it to the entity PLAYER.

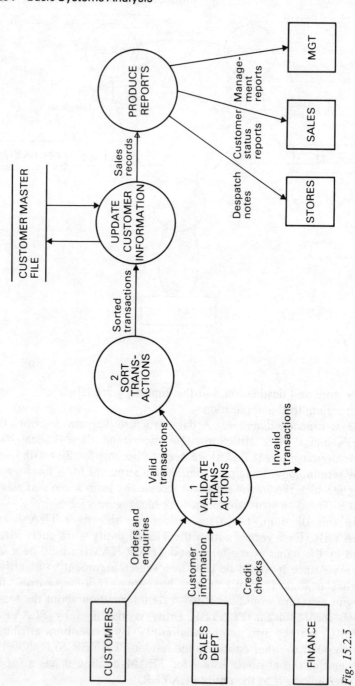

Fig. 15.2.5

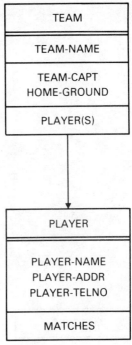

TEAM
TEAM-NAME
TEAM-CAPT HOME-GROUND
PLAYER(S)

PLAYER
PLAYER-NAME PLAYER-ADDR PLAYER-TELNO
MATCHES

Fig. 15.2.6

We could find that in carrying out our analysis, the league secretary needed to be able to identify the matches in which each PLAYER had played. The LDSD is easily expanded to show this (Figure 15.2.7).

Logical data structure diagrams do not predict the physical structure of the data base. They show the entities that make up the data base and the attributes that identify and describe the information within an entity. Logical pointers link the entities to show their interrelatedness.

Data dictionary   In Chapter 2 we discussed the importance of recording all the characteristics of data as part of systems investigation and analysis. We said that each item of data should be uniquely identified and defined and that information about data items can usefully be collected together into a data dictionary. When using structured methods, data dictionaries are essential because they record the data, dataflows and cross-references for the complete system.

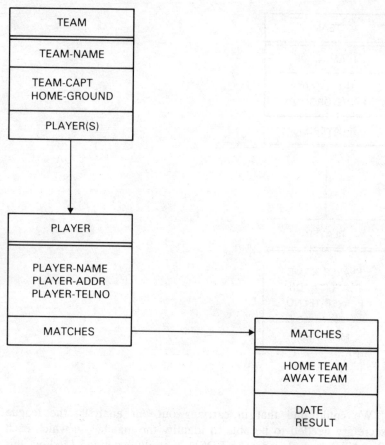

Fig. 15.2.7

Data dictionaries help to achieve the following desirable objectives:

(1) A standard definition of all terms used in a system
(2) Easy cross-referencing between subsystems, programs and modules
(3) Simpler program maintenance

Although data dictionaries are usually thought of as containing data about data, with structured methods they can conveniently record information about dataflow diagrams, as well. In this case, the data dictionary reference would show the dataflow name, a picture of the source, process, data store or destination, a concise description of the data flow and information about the sources, data stores and processes shown in the picture.

Structured text   Structured text is a way of writing procedure descriptions using a limited set of clearly defined expressions to describe the logic of the procedure. It uses a subset of the English language and is sometimes referred to as pseudo-code or program design language since it describes the lowest level of process in the lowest level data flow diagram.

The constructs used in structured text are based on those of structured programming. Imperative statements represent the 'sequence', decision statements represent 'selection' and repetition statements represent 'iteration'. The general rules for preparing structured text are:

(1) Be concise. Don't ramble and don't use 'however', 'but', 'unless'.

(2) Use verbs that work. Don't say 'do' or 'process', say 'send', 'give', 'add' etc.

(3) Document all specific system nouns in the data dictionary.

(4) Avoid using adverbs and adjectives.

The following example shows how structured English can be used. Let's begin with the following piece of narrative:

When the customer has ordered over £10 000 worth of goods this year, and the size of this order is over £500, allow a discount of 5 per cent. If, however, the customer's orders to date are less than £10 000, but this order is over £500 then allow 2 per cent. If this order is less than £500 then no discount should be given except for customers whose orders to date come to more than £10 000, in which case a discount of 3 per cent can be given.

In structured text this would look as follows:

IF   ORDERS-TO-DATE exceed £10 000 and ORDER VALUE exceeds £500, add 5 to DISCOUNT-ALLOWED

ELSE

IF   ORDERS-TO-DATE exceed £10 000 and ORDER VALUE exceeds £500, add 2 to DISCOUNT-ALLOWED

ELSE

IF ORDERS-TO-DATE exceed £10 000 and ORDER VALUE is less than £500, add 3 to DISCOUNT-ALLOWED

ELSE No discount given.

15.3 Structured design

We need to begin our review of structured design by recalling some of the things we know about structures. Since we've used organization charts in Chapter 2 already, we'll use them again. Let's look at the two following organization *structure* charts in Figures 15.3.1 and 15.3.2.

In both of these charts we can identify unusual structural relationships which make us think that things aren't as well organized as they might be. In Chart A, we have one foreman reporting to a superintendent, who in turn reports to the plant manager. Why does the sales promotion manager work for the manager of administration? In Chart B, we see a general sales manager heading for a breakdown by trying to control an impossibly large structure with no apparent organization in it at all. We can therefore make some structural criticisms of these organization charts/systems. Just looking at Chart B we see an excessively complex module at the top level. It probably has too many loops and decisions and very

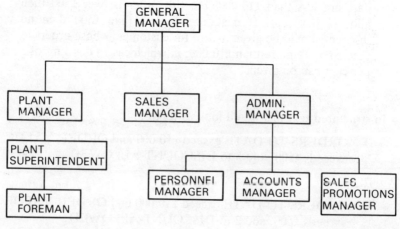

Fig. 15.3.1 Chart A

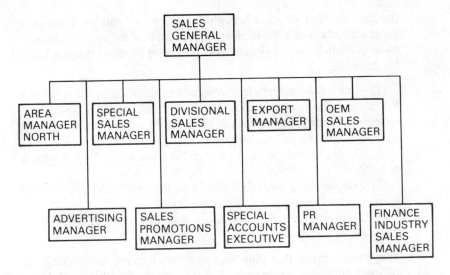

Fig. 15.3.2 Chart B

complex logic. Both charts suffer from poor 'top-down design'. There has been no strategy to break down the large complex top level problems into smaller less complex ones, continuing this functional decomposition until the original problem has been expressed as a combination of many small *solvable* problems.

Structured design strategies deal with the definition of these smaller modules and identify those characteristics which lead to the efficient arrangement of these modules within the overall design. To take just one characteristic we have identified already from the organization structure charts, span of control or 'fan out' is an important indicator of good design: more than ten or less than two subordinate modules from one superordinate module usually means that there's a better way of doing it! There is further discussion of structured design strategies in our follow-up book, *Practical Systems Design* (Pitman Publishing, 1984).

Structured walkthroughs Structured methods in general and structured design especially give particular importance to the review of analysis, designs and programs. These reviews are carried out through formal meetings to identify errors, omissions and ambiguities in any of the documentation produced during the systems development process. These review meetings are called structured walk-throughs.

We can identify several different kinds of walkthrough and although the idea of walkthroughs began with 'walking through a program', the same principles can be applied to walking through most outputs from the system's development process. We can have, for example:

(1) *Analysis walkthroughs* Where the primary purpose is to look for inaccuracies, ambiguities and errors in the analysis of an existing system. Attending such a walkthrough would be the user(s), systems analyst(s) and the project manager.

(2) *Design walkthroughs* To look for flaws in the logical system design.

(3) *Programming walkthroughs* Usually attended only by programmers to identify problems in the code.

(4) *Test walkthroughs* To ensure that the test data will test all the parts of the system.

This doesn't mean that only four walkthroughs are held during the system's development process. The four kinds of walkthrough listed may take place many times—for each program, for each part of the design, and so on.

What do we expect a walkthrough to achieve? Firstly, as will be apparent already, we expect to find mistakes of a variety of kinds: errors, omissions, contradictions, logical errors, inconsistencies, etc. Walkthroughs are very successful in finding errors and in some cases there is a dramatic decrease in the number of errors which subsequently come to light during production running. It is important to say at this point that walkthroughs are not held to correct errors. Putting right the faults that have been found is done by the analyst or programmer. Walkthroughs are not an attempt to introduce analysis and design by committee.

We also get a number of less tangible but nonetheless observed and desirable benefits from using walkthroughs. In the first place, it gives us the opportunity to involve users in the identification of problems in the analysis, and shows that while subjecting their operations to a critical appraisal we are not afraid to invite them to do the same to us. Secondly, the very nature of walkthroughs encourages analysts and programmers to improve the quality of what they are doing. This applies equally to systems designs, programs, test plans, course designs, user manuals and everything else in systems development. None of us like to show up shoddy work to our colleagues. Thirdly, walkthroughs improve the competence of the junior staff participating in them because they learn from the good practice which is revealed by a walkthrough.

A variety of people may attend a walkthrough, but three roles are essential: these are 'chairman', 'recorder', and 'author'. The chairman is usually the project manager or perhaps the programming team leader. His job is to focus attention on the objective of the review and keep unnecessary discussion to a minimum. Experience shows that walkthroughs lasting longer than two hours have been allowed to run on too long. The recorder's role is simple: it is to record the action list of all points on which further action is necessary and note whose job it is to clear them up. This then becomes the post-walkthrough task list. The author's job is to take everyone through the subject under review. There is no need for these three roles to be played by different people. The project manager could easily be both chairman and recorder, for example. Conducting walkthroughs properly isn't easy but it's not all that difficult either. Below are some suggestions which have been found to work in practice.

Before the walkthrough the project manager decides who should attend and who should play the roles listed above. Other people attending will know what the work under review is supposed to achieve, where it connects with other parts of the system and some of the real life problems it will have to handle. The time for the review should be properly scheduled so that everyone can attend and should be held in a convenient, quiet, private meeting room. The documentation which will form the basis of the review should be circulated well beforehand. Structured walkthroughs don't work properly unless *everyone* attending prepares for them. Remember also that structured walkthroughs take place by consent; in other words, when an author agrees with his project manager that there is something to review.

During the walkthrough the chairman keeps control of the meeting, making sure that everyone contributes and that a proper action list is prepared. The author walks people through the work in general and then in detail, piece by piece. The reviewing audience may often try out particular cases to see if the work under review can handle them. Disagreements are resolved by the chairman and sometimes walkthroughs are adjourned until another day if major problems have been identified and need to be resolved before work can proceed. It's important to keep proper notes during a walkthrough. Even though the objective is to find errors and not develop solutions, it would be foolish to ignore the good ideas which arise spontaneously from these meetings.

Inevitably, there are human problems when holding walkthroughs.

Often these are ego problems; none of us really like our work taken to pieces, yet most of us relish the opportunity to do it to someone else's work. Typical walkthrough problems can be grouped under the following headings:

(1) *Attitude problems* These all relate to reluctance to participate and various reasons can be advanced for non-participation. The underlying reason may be the plain fear of having one's own work exposed to criticism or even the fear of criticizing the work of others. Since the principle behind structured methods — and behind walk throughs in particular — includes the idea of 'egoless' design, it is well worth providing training and counselling in transactional analysis and team building to overcome these attitude problems.

(2) *Lack of commitment* Successful walkthroughs require commitment from everyone participating in them. People who don't try hard enough are effectively reducing the quality of the system.

(3) *Using walkthroughs improperly* Problems here are usually caused by trying to achieve too much at a walkthrough — remember the time limit of two hours. Also, people sometimes think that walkthroughs are used to judge their performance, with the consequence that more faults found in their work equals less salary. This would be far too simplistic a way to appraise analysts and programmers, who are required to possess a range of skills, abilities and knowledge. Perhaps the solution is to keep the author's manager out of the walkthrough.

So structured methods are a set of procedures to help in the development and maintenance of new systems. Structured methodology is characterized by the use of graphic documentation and by the separation of the design phase into logical design and physical design. It aims to help the analyst and user to work together more effectively to produce good systems, and, through the use of graphic documentation, to describe the system in unambiguous terms so that it can be understood fully by the users and programmed correctly.

15.4 An example of a complete structured design method

The following example of a complete structured design method is the 'Structured Systems Analysis and Design Method' (SSADM) as used in UK central government.

SSADM breaks down the systems development process into six phases with each phase further divided into steps and activities.

There are clearly defined interfaces between each step in the form of working documents, and criteria are established for review and project reporting. The six development phases are:

(1) The analysis of the current system
(2) The specification of the required system
(3) The selection of user service levels
(4) Detailed data design
(5) Detailed procedure design
(6) Physical design control

The principles on which SSADM has been constructed include many of those discussed earlier in this chapter. Specifically, SSADM states that:

(1) Development work should be based on tangibles such as agreed report formats rather than a broad statement of requirements.

(2) Project development staff should be given detailed rules and guidelines so that the next activity to be carried out is always carried out.

(3) Logical design should be separated from physical design, and as much design work as possible should be done on paper before there is any commitment to implement it.

(4) Design assumptions should be clearly stated and the installed system must feed back actual figures to compare against assumptions.

(5) Data structures should determine system structures.

The overall methodology includes a set of techniques and logical tools with rules and guidelines for their use. Standard forms are provided for development documentation and an overall framework defines the interfaces, review points and content of working documents.

Data structures play an important role in SSADM and indeed, SSADM is a data-driven method of systems analysis and design. Three important reasons are put forward in support of this approach. Firstly, data structures which reflect the natural relationships among business entities provide a stable picture for defining information requirements. Secondly, systems built around data structures are less designer dependent and more objective than those defined by function, and, finally, users see systems in terms of the data that the systems maintain, and new requirements are expressed in data terms. SSADM views data in three ways: as entity models (logical data structures); dataflow diagrams and entity life histories.

The entity model is the user's view of the business, describing the data required and the relationships between them. It is based on logical data structures and is used as a basis for developing the systems which support the business. Dataflow diagrams show how data moves around an information system and how data moves between the system and the real world. The third view of the data is provided by the life history diagram which shows the effects of time on the system. This shows all the events that affect an entity and the order in which they occur, including creation, modification and deletion.

The relationship between these three views of data is important. Dataflow diagrams show the day-to-day running of the system and how processes, transactions and system-maintained data fit together. They also show how the system links with the real world. The entity model shows the relationships that are represented in the data of the system and how this data can be accessed, and each entity in the model should be recognisable as a data store or part of a data store in the data flow diagram. The entity life history shows how entities change over time, the events the system has to deal with and the order in which they occur for each entity. Taken in combination, these three views provide a comprehensive picture of how the system uses and organises data.

Like all structured methodologies, SSADM works in a project management framework. It breaks down the systems development process into six stages, each of which is broken down further into steps, which are then themselves divided into tasks. There are clearly defined interfaces between each step in the form of working documents, and criteria are established for review and project reporting. The first three stages are concerned with analysis, the last three are concerned with design.

(1) *The analysis of the current system.* This is concerned with building a detailed representation of the current physical system and identifying its problems and shortcomings.

(2) *Specification of the required system.* Physical constraints and restrictions are removed to leave a set of logical processes and data structures which can satisfy the needs met by the current system. The ideal logical view of the current system, as developed during analysis, is extended to include the new services required by the

users. This is then expanded into a detailed specification of the required system but still concentrating on *what* it should do rather than *how* it should do it. From this stage a number of business options will be defined, and the user will select one.

(3) *Selection of the technical option.* From the detailed specification at stage 2 of the business option, a number of implementation options are possible. These options have different characteristics and offer different levels of service to the user in terms of response time, flexibility, etc. The preferred option is then chosen.

(4) and (5) *Data and process design.* From the specification of the selected option these two stages are concerned with the detailed logical design of the new system. They are concerned respectively with the design of the data and the processes within the system, but they interact and are necessarily carried out in parallel.

(6) *Physical design.* Here a rough first cut database design is prepared to enable rough cut program outlines to be developed. This, however, is not an optimum solution and, consequently, is put through an iterative process to ensure that the system meets its performance and storage objectives. This also produces the detailed data and process design, and implementation requirements.

15.4.1 Stage 1 — The analysis of the current system

SSADM is a two-pass analysis method. First, a feasibility study is made which incorporates stages 1–3 at an overview level. Then the output from this is used as input for detailed analysis, which is a repeat of stages 1–3 but in sufficiently greater detail for a design stage to follow.

A project is initiated by a start-up feasibility study document which describes the scope of the project, the boundary of the system to be analysed, and a detailed plan of the analysis phase. Three products are developed in parallel to describe the physical system.

1 *Current physical dataflow diagrams (DFDs)*

The initial investigation provides an overall view of the main scope of the system, the main documents coming into and going out of the system, where data comes from and goes to, and where information and data are permanently stored.

Dataflow diagrams use the following five basic symbols:

The external world entity

This is an oval shape representing an entity external to the system. It is used to represent an entity that is a source or a receiver of a document or information flow from the current system.

Flow of physical resource

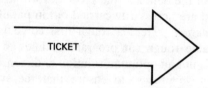

This broad arrow represents a flow of resource—typically a document or a service—with a brief description of the information being conveyed.

Process

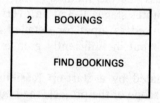

This box represents a process that manipulates information, using inputs and producing outputs. On the top left is an identification number that identifies the process. A description is also included to aid understanding.

Data store

This symbol represents a store of data. It may be a computer file or a manual file or an in-tray. In the required system a data store is identified by a reference number starting with a D, but in the current system all manual data stores start with an M.

Data flows

TICKETS

This symbol shows the flow of information between processes or between a process and an external entity. Often the flow of information is described to increase understanding.

Initially, an outline DFD is produced which shows the system boundary, the document and information flows between the system and other departments and also outside agents.

To produce an outline DFD, the following steps should be followed. Initially, a list of the physical documents involved is made, i.e. booking, invoice, ticket. The diagram is then drawn, with the document and information flows and the external entities. Finally, the document is checked for completeness with the user. Figure 15.4.1 shows an outline DFD for a holiday booking system.

A first level data flow diagram is produced by expanding the

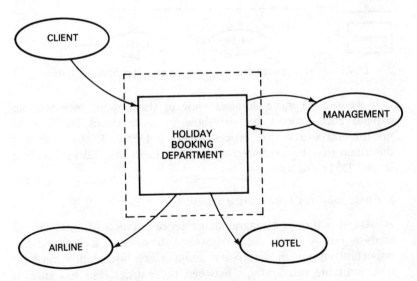

Fig. 15.4.1 Outline DFD for a holiday booking system

system within the boundary and adding the processes, information and document flows, and data stores. This is achieved by listing the processes to be carried out in each document, and including them on the diagram. If a large number are identified—more than 16—then these should be combined in terms of time and/or function. Figure 15.4.2 is an example 1st level dataflow diagram for a holiday booking system.

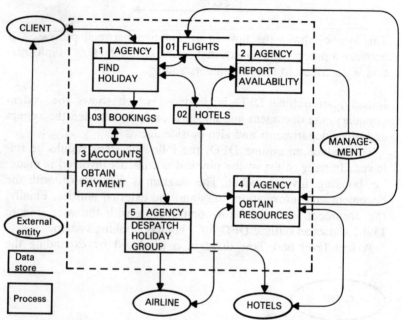

Fig. 15.4.2   First level dataflow diagram for a holiday booking system

If we need a more detailed view of the system, then we can expand each process box to show in more detail the flow of information within that process (level 2 DFD). If the processes described may be expanded again, then these may be shown on a level 3 DFD and so on.

2 Entity models of the current system

A data structure is a diagrammatic representation of data. As an analysis tool it allows the analyst and the user to understand the important entities in the system about which information must be held, and the relationships between the entities. For example, a holiday booking usually has associated with it two flights, a booking at a hotel, etc. It is important at a very early stage to understand the information requirements of the system, so that it can be clarified and agreed with the user before proceeding. Entity types are difficult to define in detail. They are not individual data items, but represent things of significance, about which data must be held, such as hotels in the holiday example. These are represented by rectangular boxes, with an entity description included.

Relationships between entities are indicated by a single line between the boxes. In such a relationship as between booking and flights, one booking may be associated with many flights. A crow's foot indicates which entity is the one — the *master*, and which is the many — the *detail*. For example:

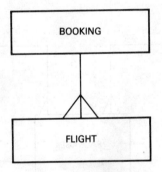

In certain cases, the relationship is not one to many because an airport may cater for many flights. Similarly, a flight may touch down at a number of airports on its journey. This is known as a many-to-many relationship and is shown as follows:

These many-to-many relationships can be easily misinterpreted and to make them clearer a link or junction box is used to convert the many-to-many relationships into *two* one-to-many relationships like this.

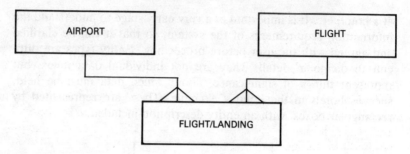

Here an airport can now have many landings and a flight may have one or more landings.

Often a relationship is one-to-one, particularly in the early development of data structures. If this is the case, it is usual to combine the two entities in the relationship into one.

The following logical data structure (Fig. 15.4.3) has been produced for the holiday booking system.

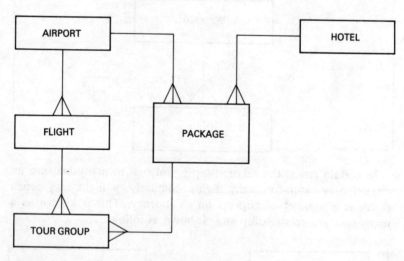

Fig. 15.4.3 Entity model for a holiday booking system

3 Problems/requirements list

No business will remain static in the way in which it operates. The environment and the operation of the business are completely dynamic and constantly changing. The difficulty with manual and computer systems is that they tend to be static and rather rigid in the way in which they operate. In time they no longer serve the business efficiently, and management decide that a more relevant system is required.

The problems/requirements list allows the systems analyst to identify where the operation of the current system is unsatisfactory, and which additional features the users would like to see incorporated into the new system.

The problems/requirements list will contain a problem reference, the reference of the person who raised the problem, its priority, the problem itself and the suggested solution (see Fig. 15.4.4).

Problems/requirements list for a hotel booking system					
Ref. No.	User Ref.	Priority	Problem/ requirement	Possible solution	DFD/LDS Ref.
BO3	H.M.	3	Problems with identifying hotel availability in a reasonable time scale	Provide on-line enquiry facility for hotel availability	D2
BO4	J.J.	7	Difficulty with understanding the proliferation of different booking forms	Provide a common input booking screen for all holiday groups	D5

Fig. 15.4.4 Problems/requirements list for a hotel booking system

At the end of this stage, the dataflow diagrams, entity models and problems/requirements lists are formally reviewed with the user to ensure their accuracy and completeness before proceeding to the second stage.

15.4.2 Stage 2—Specification of requirements

For the first stage, the current physical system is described in terms of DFDs and logical data structures. The DFDs in particular reflect the physical attributes of the current system.

The specification of requirements stage consists of a number of steps. Initially the physical constraints imposed by the way the system is organised are removed and a model of the required system is then produced by the analyst modifying the logical DFDs and the logical data structures to meet the requirements specified in the problem/requirements list. At this stage a catalogue of functions and a description of the specified entities are built up. The third view of the data, the entity life histories, is then built up for each entity on the logical data structure.

Converting physical to logical DFDs

One of the major problems with the development of DFDs in stage 1 is that they describe the way that the current system is implemented, rather than describing current user policy. The current DFDs will contain processes that occur for political, historical or mechanistic reasons. Similarly, for implementation reasons, data stores may have been viewed in a number of different ways. Unfortunately, there are no general rules that can be applied for the rationalisation of processes. Data stores however can be rationalised in two ways. If there are two or more main data stores and they have identical key data items, it is possible to combine them into a single data store. Similarly, temporary or transient data stores may appear because the method of working in the current system requires the information to be stored for a short while between processes. This may not be logically necessary, and should be removed from the DFD if it is not needed.

Logical data structures (LDS) and logical DFDs

Once the DFDs of the current system have been drawn, the analyst needs to create a model of the required system based on the LDS, current logical DFDs, and the problems/requirements list. Typically, this is an iterative process, although the LDS is generally examined first. Once this has been done, a set of business options should be produced and reviewed with the user.

The operations to create the required LDS are very similar to those associated with creating the initial LDS. Usually the current structure will be modified for changes, or new features added. Provided that the requirements for the new system are not radically different from the existing one, then the changes are unlikely to be too extensive and are likely to be carried out by inspection, and with reference to the required DFDs.

Once the required LDS has been produced then the DFDs can be amended, or added to depending on the problems/requirements list. One area to be considered is the meeting of system requirements where they are currently felt to be inadequate. Also, additions are required when services required are not currently available. Finally, the analyst should ensure that the business and systems objectives will be met.

Creation of entity life histories
When the required logical data structure is produced, it provides a view of the data required by the system and the relationships between the individual entities. Similarly, the required DFDs provide a view of the flow of data through the system. The third view, the entity life history, helps to tie the information together to show what happens to an entity over a period of time when it is affected by a series of events. In this context, events may be regarded as stimuli which initiate processing within the system such that the state of an entity or entities may be changed in some way, for example, the creation, amendment or deletion of an occurrence of that entity. Entity life histories should also take into acocunt non-events such as the non-arrival of payment for a booked holiday. Events for a particular entity may be described in a hierarchical diagrammatic fashion, in one or more of the following three ways:

1 Sequence of events

Here events are read downwards from the left, and then from the left to right. In this case, events *must* occur in the order presented. For example:

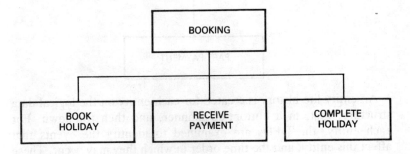

2 Selection of events

A selection shows a number of options, where an occurrence may be at any one time in only one of the instances. The optional instances are shown with a circle in the top right hand corner. One of the options *must* be selected. If it is possible that no instance may occur, then a null instance must be shown. For example, a holiday-maker may decide to book a first-class or a club-class seat or decide to make other travel arrangements and hence not need a flight at all:

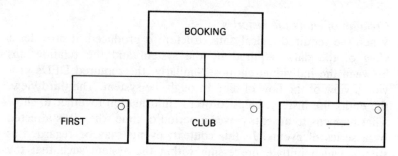

3 Iteration of events

Here a number of identical events occur over a period of time, for example part-payments for a holiday to be made once per month. This is shown by the event box containing an asterisk in the top left hand corner, for example:

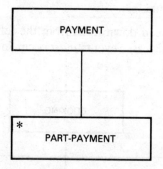

An entity life history is created for each entity on the logical data structure; first, in a bottom-up manner, and then top-down. For each entity, the DFDs are examined to identify the events that affect this entity, and the time order in which they may occur. These are initially included into an event/entity matrix, from which the ELHs are built up. As this process is being carried out, a number of errors/omissions in the LDS and DFDs may come to light and will need correcting.

For example, the diagram opposite, on p. 275, shows a partial entity life history for the booking entity in a holiday booking system.

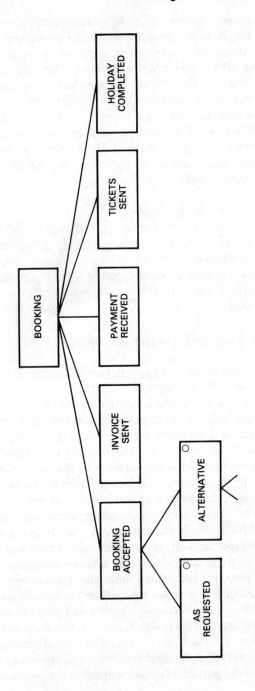

Creation of event, function and retrieval catalogues
The entity life histories produce an indication of the events that occur for a particular entity. At the end of the step, the analyst produces the ELHs and event catalogue. The catalogue records events and details of the entities that are changed by the events.

Whereas the event catalogue is produced from the ELHs, the function and retrievals catalogues are produced from the DFDs of the required system. The function catalogue records update functions together with description and DFD reference number. The retrievals catalogue records all processes on the physical DFDs which are retrievals only

Creation of logical dialogue outlines
From the function and event catalogues above, overall logical dialogue outlines are produced for the on-line update in the system.

The documentation produced in the process of specifying the required system is then brought together in a requirements specification, which is formally reviewed with the user before proceeding to the third stage.

15.4.3 Stage 3—Selection of technical options

In stage 2 of the methodology, a detailed analysis of the required system is produced, and agreed with the user. There may be a number of methods of implementation, batch or real-time for example, there will be cost, performance and time implications. Stage 3 defines a framework for the design stages that follow. A number of skeleton options are drawn up by the analyst, and then presented to the user. After discussion with the user, an option, or a combination of the options, is selected for implementation. At the end of this stage, the selected option is specified in more detail with a formal system specification produced for agreement with the user. Once each option has been outlined, it may be expanded into a more detailed specification. Information will include technical hardware and software details and enough additional information for the user to be able to understand how the system will work, the significant design features and the outline costs. Information about reliability, fallback and recovery, security and possible maintenance and expansion capabilities is also necessary. Since the decision to be taken will also take account of financial consideration, it is also important to include a detailed cost/benefit analysis and some indication of how the particular option will affect the user environment.

Once the options have been defined, meetings should be arranged with the user and user's management where the possible options are presented and reviewed, with one or a combination of the options selected. Once the required option has been selected, the last step is to produce a detailed specification which will provide a basis for the following design stages. Here the user option will be expanded to show any changes to the DFDs, LDS or ELHs produced at stage 2, but needing modification in light of the user option selected. Also included should be a detailed development plan for the work required, and the overall performance objectives in terms of response time and batch turnround time together with the associated volumes.

15.4.4 Stages 4 and 5 — Data and process design

Once the user option has been selected, then the analysis phase is complete and the design process can begin. It is important that, for as long as possible, the design should consider only logical design with a detailed logical model being produced. It is at stage 6 that the implications of the choice of hardware and software, data management system and so on should be considered.

Stages 4 and 5 are carried out in parallel, and interact at all times. Stage 4 looks at the data design, and is concerned with developing two products: optimised third normal form (TNF) relations, and a composite logical data model (CLDD). Stage 5 develops more detailed process outlines.

Third normal form data analysis
Data analysis is the process of discovering and recording all the data needed in the system and showing its detailed structure. As we saw in Section 2.8, Codd's 'relational theory of data' showed in mathematical terms how data may be described and the relationship of each part with the rest. This has been developed into a set of comprehensive rules so that the data so defined is in its simplest form allowing maximum flexibility of use and making the overall structure of the data clear. TNF describes those rules to ensure that normalised data structures are produced, and that a complete understanding of the required data is gained.

TNF data analysis applies the TNF rules to each of the input/ouput descriptions, entity descriptions and report formats from stage 2 to produce an optimised set of TNF relationships. These relationships are then converted to a generically correct data structure by using the same diagrammatic techniques as for logical data structures, following a relatively simple set of rules.

Composite logical data design (CLDD)
Once the TNF analysis has been completed, then two data models are available, from the TNF process and from the logical data structure of the required system. This step combines these two documents into a single logical data model.

Logical enquiry and update process outlines
Here the update process outlines are produced to satisfy the processing requirements of the CLDD and extended to show the detail of processing required. Also at this stage, new process outlines are produced for the enquiry facilities identified during the analysis stages from the retrievals catalogue.

Also, the logical dialogue outlines produced in stage 2 will be refined further and associated logical design controls produced. The logical design control can be thought of as analogous to a lower level menu which handles a group of transactions.

15.4.5 Stage 6 — Physical design

Here the logical design created in stages 4 and 5 is taken a stage further, producing a physical design based on the hardware and the software chosen for implementation. The first step is to carry out the first-cut data design. Here a set of database-specific rules are applied to the composite logical data model to produce a data model that is implementable with that particular data management system. In parallel with this, the first-cut program design is carried out, whereby logical process outlines produced in stage 5 are converted to physical process specifications by a similar set of database-specific rules. The physical designs produced do not, however, necessarily meet timing, sizing and performance requirements laid down for the required system. An iterative 'physical design control' stage is carried out to refine the design to a level where it can be implemented to meet all the system requirements. Once an acceptable design has been achieved, then the anlayst must prepare for the implementation of the system. Program development and system test plans should be drawn up, operating instructions and manual procedures are specified, and detailed program specifications are produced.

The implementation of the system can now begin.

Note: Structured Systems Analysis and Design Methodology (SSADM) is the methodology used by the UK government, and information about it can be obtained from the Central Computer and Telecommunications Agency (CCTA), Riverwalk House, 157–161 Millbank, London SW1P 4RT, UK.

16 Case Study: SSADM in Systems Analysis

16.1 Background information

16.1.1 Introduction

COMPUTRAIN is a company offering systems, programming, operations and management training to customers in the UK, Europe and the Middle East.

The company is part of a larger information technology group and in its early days all training was provided 'in-house' to the parent company. Since then, the company has grown considerably, first in the number and range of courses that are provided, and second in the number of external customers using COMPUTRAIN's products. These 'products' may be defined as scheduled courses given at COMPUTRAIN's premises and in hotels, 'on-site' courses for some of the larger customers and 'special courses' which use one or a number of modules specifically tailored for a customer requirement.

The company also markets self-instructional material and computer books to a wide selection of customers. This is currently outside the scope of this investigation, but in the production of a new system there must be consideration that it should be enhanceable to cater for the marketing of, and revenue reporting on, these materials.

Problem
The main problem area is that the current system which handles the bookings of students on courses, invoicing on completion of the course, booking of necessary conference accommodation and allocation of resources in terms of equipment and lecturing staff, is completely manual and has changed little since it was first set up. This means that the system is unable to cope with the increase in volumes and cannot provide the control information now required by the management.

Objective
The overall objective is to investigate the current system, identify all the problems and requirements, and design a computer based system to resolve the problems and provide the appropriate information for better marketing and management control This example covers the essentials of the Analysis Phase of SSADM—stages 1–3.

16.1.2 Current system description

Bookings
Customer enquiries and bookings are received by letter, telephone and on the company booking form which is included in the course brochure sent annually to regular customers and in response to previous ad hoc enquiries.

(1) *Enquiry* On receipt of an enquiry, a sales enquiry form is completed (see Section 16.2, Form 1). The information entered is as follows:

- Date of enquiry
- Name of individual
- Job title
- Name of company
- Address
- Telephone number and extension

Entries are also made at the bottom of the form as to the action taken (e.g. information sent, provisional booking made, etc.). If this is only an enquiry, then the form is passed to one of the sales force for subsequent follow-up and an acknowledgement sent to the customer, together with a course brochure if the customer requires it.

(2) *Provisional booking* When a provisional booking is made, either by letter or telephone call from the customer or from contact with a salesman, the customer is asked to return the booking form and the sales enquiry form is marked as provisional. If the provisional booking is made for a course taking place in the next 16 weeks, the customer's name is added to the relevant course list on a large board of scheduled courses, otherwise the sales enquiry sheet is filed in date within alphabetic sequence. A copy of the annotated enquiry form is sent to sales for information.

(3) *Confirmed booking* Once the booking is confirmed, either by the customer sending in the completed company booking form,

or through sales, the enquiry form is marked as confirmed and the name on the 'board' list is ticked as confirmed. An acknowledgement of the confirmation is sent to the customer.

(4) *Joining instructions* Three weeks prior to the course starting, confirmed candidates are sent their joining instructions which include details of the course venue, start and finish dates and times, a detailed course agenda and any relevant pre-course documentation. Provisional candidates still unconfirmed are notified to sales for follow up.

(5) *Cancellations* When customers cancel a booking (usually because other commitments arise), the sales enquiry form is annotated cancelled and a copy sent to sales. If they are already booked on the 'board' list, the name is crossed out.

(6) *Customer file* A customer file is continually updated with copies of correspondence, proposals by sales, booking forms returned by the customer and joining instructions sent to the customer.

(7) *On-site/special courses* Enquiries for either on-site or special courses are entered onto the sales enquiry form as for any other enquiry and passed through to sales. Once the order has been won and the price, date and location have been agreed with the customer, and the content agreed with the lecturing staff, together with lecturer availability, this is then passed back to bookings as either a provisional or confirmed booking. The procedures are then followed as for normal scheduled bookings.

(8) *Accommodation* A register of COMPUTRAIN's own course accommodation is held by bookings and scheduled courses are allocated accommodation in discussion with management. This allocation is done every 8 weeks in advance, to reflect the board list of courses. Certain courses require hotel accommodation—both residential and non-residential. The hotel site, facilities and price are first agreed with management and then the relevant accommodation is booked. The details of the location and facilities offered by the hotel are sent out with the course joining instructions.

(9) *Monthly report* At the end of each month, a report is compiled for management, listing the courses given, scheduled or on-site, the delegates attending and the companies they came from. A course log is also kept of the number of delegates attending through the month, and totals kept.

Invoicing
Invoicing is carried out by the group's head office through a central

billing system. Head office also handles the associated cash collection and credit control. Invoice details are sent to head office from COMPUTRAIN for input to the system.

At the end of each month, all the details concerning completed courses are sent to the invoicing section in head office, and include:

- Course name and date
- Course fee with any relevant discount
- Delegate name and company

Using these details and billing name and address information from a customer list and course list, a billing request form is typed up for each customer to be invoiced by head office (see Section 16.2, Form 2). This contains:

- Customer number (from customer list)
- Course code (from course list)
- Course name
- Selling price (not including VAT)
- Cost centre code (for COMPUTRAIN)

If the customer has booked delegates on a number of courses during the month, then a block billing request form is filled out (Section 16.2, Form 3).

Once all the billing requests have been typed up, they are sent to head office for input to the main billing system. Since this is merely a data preparation function, little checking is done. Once all the data is input, then head office do a 'dummy' run of the invoice production process and these details are sent back to COMPUTRAIN for checking against the original billing request forms. Requisite changes are then made and sent to head office for correction and finally the invoices are prepared and sent off to customers.

(1) *New customer and courses*   A list of all customers and courses, together with their associated codes is produced on the group computer and sent to COMPUTRAIN monthly or on request.

When a new course (which can be defined as one of COMPUTRAIN's 'products', see Section 16.1.1) is developed and scheduled, a product set-up form is completed (see Section 16.2, Form 4), and includes:

- Product code
- Description
- Unit title

- VAT code
- Standard price

Similarly, for new customers, a customer set-up form (Section 16.2, Form 5) is completed with the following:

- Name
- Address
- Town
- County
- Postcode
- Country
- Tel. No.   Telex No.
- Contact name
- Set up date

The set up forms are batched weekly and sent to head office for the update of the billing system customer and course files and the production of updated listings for COMPUTRAIN.

(2) *Credit notes*   To cater for mistakes on invoices, or customer dissatisfaction with a course which is justified (a fairly rare occurrence), credit notes are raised.

The query/complaint is received by COMPUTRAIN either directly from the customer, or the head office credit control function. Once credit has been agreed by management, a credit note request (Section 16.2, Form 6) is completed and sent to head office for processing through the billing system. A credit note is produced and sent to the customer. The entries include:

- Customer name
- Customer number
- Reason for raising the credit note
- The total credit amount
- Authorisation signatures

(3) *Sales revenue*   At the end of each month, when the billing request forms are completed for sending to head office, a sales revenue file is updated with the following details:

- Company (to be billed)
- Name of course
- Dates
- Total price (including accommodation if residential, but not including VAT)

This information is then used to produce a report for management showing the total expected revenue for courses running in that month.

Summary
The outline of the current system highlights the need for the automation of certain booking and invoicing functions to help handle the increased volumes of information. Also, once the operational aspects of the system have been redesigned, it is essential that relevant extracts of information are provided to enhance productivity, forecasting and management control. These are specified on the Problems/Requirements lists that follow, but some examples are as follows:

- Production of chase-up lists of provisional bookings for use by sales
- Analysis of forecast against actual revenue
- Analysis of course productivity in terms of attendance and revenue by month
- Company trend analysis showing courses attended, numbers and sequence
- Analysis of course booking dates against start dates to identify optimum lead time for advertising
- Quarterly analysis, in advance, of lecturer and equipment availability for on-site requirements
- Automated production of course delegate/company lists for inclusion in the joining instructions.

Current system volumes

Customer list	:	5000 (no. of customers)
Course list	:	100 (no. of courses)
Enquires per month	:	200–250
Bookings per month	:	250–300
Invoices per month	:	250–300
Acknowledgements and course details per month	:	200–250
Joining instructions sent per month	:	250–300

16.2 Forms used in the current system

```
SALES ENQUIRY                           DATE ............

Name of Individual ..............................................

Job Title .......................................................

Name of Company .................................................

Address .........................................................

................................................................

................................................................

................................................................

Tel. No. ............................... Extn .............

Referred by:

Summary:




_____

Action:                          Booking made:

Information given/sent by  [    ]   Provisional   [    ]

Database entry made        [    ]   Confirmed     [    ]

Action taken by Sales Exec:




Returned to        (date) ...........
```

Form 1 Sales enquiry form

BILLING REQUEST FORM

Cost Centre Code   M _ _ _   Suffix   _ _

Customer No   M _ _ _ _ _   Product Identifier   _ _ _ _ _ _

Product Code   M _ _ _ _

Description   M _

Invoicing Frequency   M _   Selling Price M _ _ _ _

Contract No   M _ _ _ _ _   Purchase Order No   _ _ _ _ _

                     Zeroise Quantity after Invoicing (y/n) M _

            *Update Price or Quantity on Billing Input (p/q) M _

                                             Sales Originator _ _ _ _

Date product first used   M _ _ _ _   Date product last used M _ _ _ _

Period for first billing (yyww) _ _ _

                  If quantity, type quantity to be billed M _ _ _ _ _ _

* INVOICE CALCULATON

PRICE INVOICE: Selling Price X 1

QUANTITY INVOICE: Selling Price X
                  Quantity (X
                  multiplication) or
                  (/ Divisor)

Sig _____   Date _____

Form 2 **Billing request form**

BILLING REQUEST FORM COST CENTRE: _ _ _

CUST NO: _ _ _ _ _ _   SUFFIX _ _ _

CUST NAME: _ _ _ _ _ _ _ _ _ _ _ _ _

ADDRESS: _ _ _ _ _ _ _ _ _ _ _ _ _ _   (NOT MANDATORY)

_ _ _ _ _ _ _ _ _ _ _ _ _ _

_ _ _ _ _ _ _ _ _ _ _ _ _ _

CONTRACT NO: PURCHASE ORDER NO:

PRODUCT CODE	DESCRIPTION (UP TO 3 LINES X 50 CHARACTERS PER PRODUCT CODE)	BILL FRQ	QTY	PRICE	VALUE	Y/N*
	TOTAL					

DATE PRODUCT FIRST USED _ _/_ _/_ _   PERIOD FOR FIRST BILLING (YY WW) _ _ _ _

DATE PRODUCT LAST USED _ _/_ _/_ _

SIGNED: DATE:

*Zeroise quantity after billing

Form 3   Block billing request form

PRODUCT SETUP FORM

Product Code   – – – –

Description   –
                    –
                    –

                                              Multiplication/Divisor   – – – – – – – –

Unit Description   – – – – – – – – –

VAT Code   –   0 = 0% 1 = 15%

Standard Price   – – – – – – – – –

                                    Suppress Sale Unit printing (y/n) –

INVOICE CALCULATION

PRICE INVOICE: Selling Price X 1

QUANTITY INVOICE: Selling Price X
                    Quantity (X
                    multiplication) or
                    (/Divisor)

Sig _____   Date _____

Form 4 **Product set-up form**

CUSTOMER SETUP FORM

Name _

Address _

Town _

County _

Post Code _

Country _

Tel. No. _ _ _ _ _ _ _ _ _ _ _ _ _ _ _ _ _ _ _   Telex No. _ _ _ _ _ _ _

Contact _ _ _ _ _ _ _ _ _ _ _ _ _ _ _ _ _ _ _

Controlling
Cost Centre _ _ _

Sig _____   Date _____

Form 5 Customer set-up form

```
TO:                          DATE:

FROM:

                    CREDIT NOTE REQUEST

CUSTOMER NAME _____

CUSTOMER NUMBER _____ MVS A/C NO & CHECK DIGIT _____

DETAILS




                                         NET TOTAL £ _____


REQUESTED BY: _____ DATE: _____

PRODUCTION CONTROL: _____ DATE: _____

SALES MANAGER: _____ DATE: _____

BRANCH SALES MANAGER: _____ DATE: _____

O.M.B: _____ DATE: _____

MANAGING DIRECTOR: _____ DATE: _____
```

Form 6   Credit note request

16.3 The current COMPUTRAIN system

This has been covered in some detail by the background information and copies of current forms given in Sections 16.1 and 16.2. In the first stage of SSADM the task is to analyse the current system. Some examples of the output of this stage are:

(a) *Level 0 DFD and System Boundary* (Figure 16.3.1) This defines the data flows and the associated sources and recipients. This is used to define the system boundary.

(b) *Context DFD* (Figure 16.3.2) Once the boundary has been defined the context diagram is represented by the central box, which represents the functions to be analysed and the data flows to/from external sources and recipients.

(c) *Level 1 DFD* (Figure 16.3.3) A diagrammatic representation of the sources/recipients of data, processes, associated data stores, and data flows in the current system.

(d) *Level 2 DFD* (p. 296) A level 1 process broken down into more detailed activities.

(e) *Current LDS* (Figure 16.3.4) A logical data structure of the relevant data in the current system.

(f) *Data Store/Entity Cross Reference* (pp. 298–299) A cross reference between the data in the LDS and the data held in data stores on the DFDs. This ensures that all the data is represented in the two views of the system.

(g) *Problem/Requirements List* (pp. 300–304) A list of the problems and requirements of the current system. In the example given solutions are not entered. These are added in stage 2 of SSADM— Definition of the Required System.

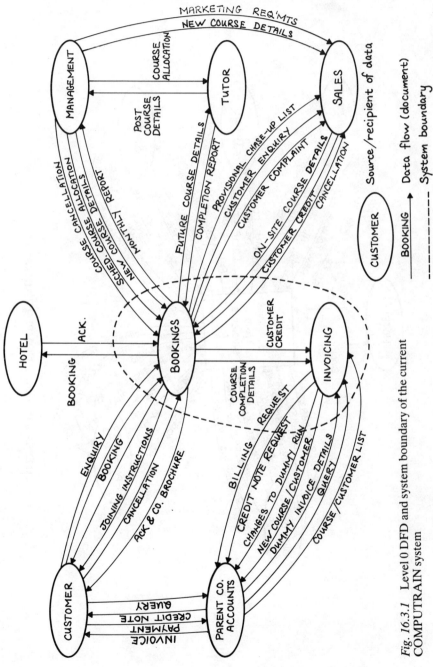

Fig. 16.3.1 Level 0 DFD and system boundary of the current COMPUTRAIN system

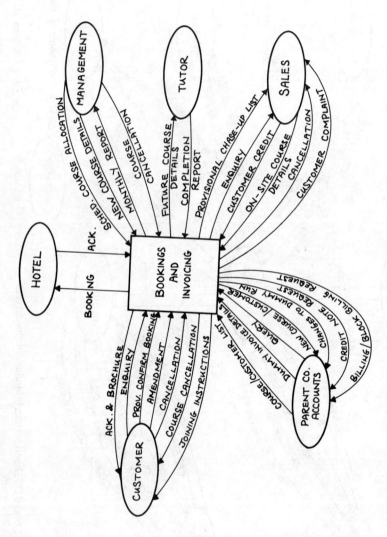

Fig. 16.3.2 Context DFD of the current COMPUTRAIN system

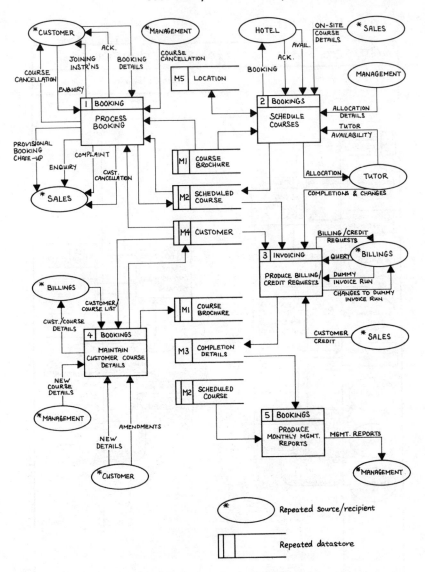

Fig. 16.3.3   Level 1 DFD of the current COMPUTRAIN system

Lower Level DFD

SYSTEM: BOOKINGS	DATE 25/ 3 /87 AUTHOR: DG

Current /~~Physical~~   Logical /~~Required~~

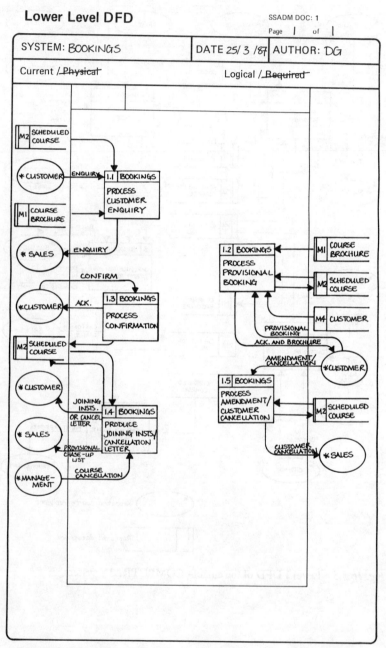

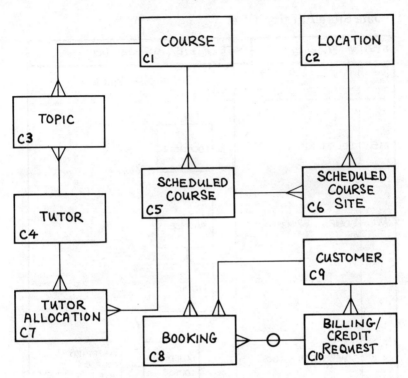

Fig. 16.3.4   LDS of the current COMPUTRAIN system

Data Store / Entity X-Ref

SYSTEM: BOOKINGS	DATE 25/ 3 /87	AUTHOR: DG

Data Store Ref.	Name	LDS Entities & Structure
M5	LOCATION	LOCATION
MI	COURSE BROCHURE	COURSE TOPIC
M2	SCHEDULED COURSE	SCHEDULED COURSE — SCHEDULED COURSE SITE TUTOR ALLOCATION BOOKING
M4	CUSTOMER	CUSTOMER

Data Store / Entity X-Ref

SSADM DOC: 3
Page 2 of 2

SYSTEM: BOOKINGS	DATE 25/ 3 /87	AUTHOR: DG

Data Store Ref.	Name	LDS Entities & Structure
M3	COMPLETION DETAILS	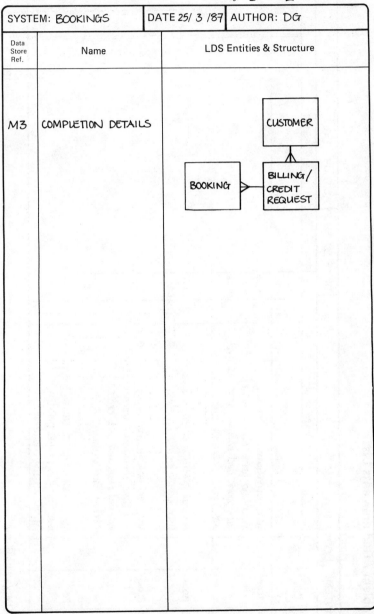

Problem / Requirements List

SSADM DOC: 2

Page 1 of 5

SYSTEM: BOOKINGS

DATE 25/ 3 /87 AUTHOR: DG

P/R Ref No.	User Name/ Ref	Prio- rity	Problem Requirements			Solutions		
			Brief Description	Ext Nar Ref	DFD/ LDS Ref	Brief Description	Ext Nar Ref	DFD/ LDS Ref
P001	BOOK	1	Increased volumes of bookings incurring large ammounts of paperwork and manual look-up and cross-referencing of files.		1.3, 1.2 1.5/ c5, c8			
P002	BOOK	1	Manual preparation of joining instructions for all courses.		1.4/ c8			
P003	BOOK	1	Any necessary correction of errors is very time consuming because of manual matching and cross-referencing method.					
P004	BOOK	1	The overall workload leaves less and less time for the prime priority - satisfying the customer.					

SSADM Version 3 Issue 1

Jan 1986

Problem / Requirements List

SSADM DOC: 2

Page 2 of 5

SYSTEM: BOOKINGS

DATE 25/ 3 /87 AUTHOR: DG

P/R Ref No.	User Name/ Ref	Prio- rity	Problem Requirements			Solutions		
			Brief Description	Ext Nar Ref	DFD/ LDS Ref	Brief Description	Ext Nar Ref	DFD/ LDS Ref
R/01	BOOK	1	Up-to-date database of customers and courses to be held at Computrain. This to supercede parent company listings.					
R/02	BOOK	1	On-line enquiry and update facilities for provisional and confirmed bookings together with amendments and cancellations.					
R/03	BOOK	1	Automatic print of joining instructions for course delegates 3 weeks prior to the start of the course.					

SSADM Version 3 Issue 1

Jan 1986

Problem / Requirements List

SSADM DOC: 2

Page 3 of 5

SYSTEM: BOOKINGS						DATE 25/ 3 /87	AUTHOR: DG			
			Problem Requirements				Solutions			
P/R Ref No.	User Name/ Ref	Prio-rity	Brief Description	Ext Nar Ref	DFD/ LDS Ref		Brief Description	Ext Nar Ref	DFD/ LDS Ref	
R004	BOOK	2	Automatic production of monthly management reports and associated batch enquiry facilities.							
P005	INV	1	Delay in update by parent company of customers and courses. This results in duplicate set-up forms being completed.							
P006	INV	1	Unnecessary duplication of effort in the preparation of invoices. Manual preparation at Computram, dummy-run of input on mainframe, check against originals, resubmission of corrections and final invoice print from mainframe.							

Jan 1986

Problem / Requirements List

SYSTEM: BOOKINGS

DATE 25/ 3 /87 AUTHOR: DG

P/R Ref No.	User Name/ Ref	Prio- rity	Problem Requirements			Solutions		
			Brief Description	Ext Nar Ref	DFD/ LDS Ref	Brief Description	Ext Nar Ref	DFD/ LDS Ref
P007	INV	1	Errors created by pressure of work during the month and these compounded by the monthly invoice deadline.					
R005	INV	1	Automatic production of customer and course lists on an ad-hoc basis from local Computrain database.		/C1, C2			
R006	INV	1	Automatic monthly production of billing / credit note requests for sending to the parent company.		/C10			

Problem / Requirements List

SSADM DOC: 2

Page 5 of 5

SYSTEM: BOOKINGS

DATE 25/ 3 /87 **AUTHOR:** DG

P/R Ref No.	User Name/Ref	Prio-rity	Problem Requirements Brief Description	Ext Nar Ref	DFD/ LDS Ref	Solutions Brief Description	Ext Nar Ref	DFD/ LDS Ref
R007	MGMT	1	Input of all resource availability (tutor and machine) and automatic scheduling facilities together with exception reporting and on-line re-scheduling facilities.		/C4, C7			
R008	MGMT	1	Production of monthly statistics. e.g. (a) Revenue: Actual v. Forecast. (b) Revenue by Tutors. (c) Revenue by Course Type.					

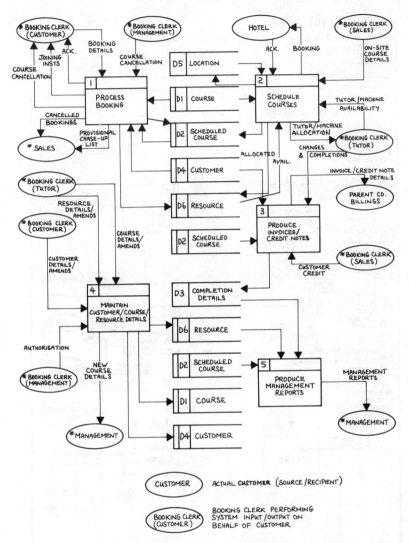

Fig. 16.3.5 Level 1 DFD of the required COMPUTRAIN system

Lower Level DFD

SYSTEM: BOOKINGS   DATE 25/ 3 /87 AUTHOR: DG

~~Current / Physical~~   Logical / Required

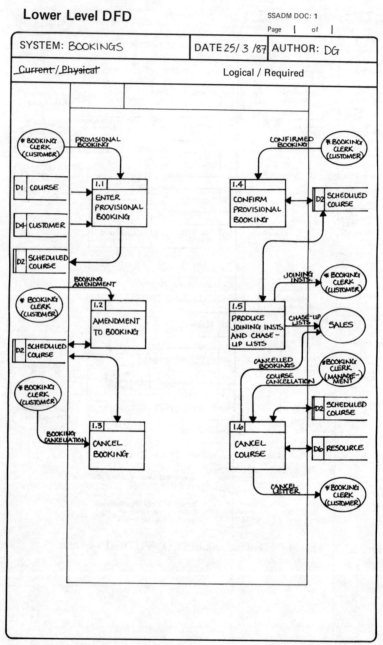

Elementary Function Description

SYSTEM: BOOKINGS	DATE 25/ 3 /87	AUTHOR: DG

Process ID	Process Name	Brief Description
1.	PROCESS BOOKINGS / AMENDMENTS / CANCELLATIONS	
1.1	ENTER PROVISIONAL BOOKING	The provisional booking is checked against course and customer and then a provisional booking is created.
1.2	AMENDMENT TO BOOKING	On receipt of a booking amendment from the customer, a relevant amendment (e.g. delegate name and position) is made to either a provisional or confirmed booking.
1.3	CANCEL BOOKING	Removes details of provisional or confirmed booking from the relevant scheduled course.
1.4	CONFIRM PROVISIONAL BOOKING	Changes the status of the booking on the scheduled course from provisional to confirmed.
1.5	PRODUCE JOINING INSTRUCTIONS AND CHASE-UP LISTS	3 weeks prior to the start of a scheduled course, joining instructions are printed for all confirmed bookings. Any bookings which are still provisional are listed for sales personnel to chase up.

308 Basic Systems Analysis

Elementary Function Description

SYSTEM: BOOKINGS		DATE 25/ 3 /87	AUTHOR: DG
Process ID	Process Name		Brief Description
1.6	CANCEL COURSE		If there are insufficient bookings on a course to make it viable, management will cancel it. This removes that scheduled course and the booking details, after the latter have been listed for the sales personnel to deal with. Resources in terms of tutors and machines are de-allocated from the scheduled course.

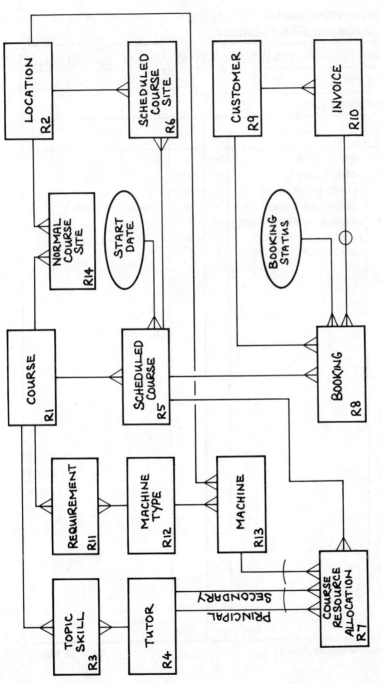

Fig. 16.3.6 LDS of the required COMPUTRAIN system

Entity Description / Optimised TNF Relations

SSADM DOC: **8**

Page I of 14

SYSTEM: BOOKINGS	DATE 25/ 3 /87	AUTHOR: DG	Del as Required

LDS | TNF | CLDD

Entity ID & Name: COURSE	Volumetrics: Ave 50 Max 100

Entity Description:

Key P/F	Data Items	For mat	Length	Comments
✓	COURSE CODE	X	5	
	COURSE NAME	X	30	
	COURSE DESCRIPTION	X	256	
	COURSE STRUCTURE	X	256	(MAIN HEADINGS)
*	NORMAL LOCATION CODE	X	5	
	NO. TUTORS REQUIRED	9	2	
	NO. MACHINES REQUIRED	9	2	
	Total Size		556	

SSADM Version 3 Issue 1

Jan 1986

Entity Description / Optimised TNF Relations

SSADM DOC: 8

Page 2 of 14

SYSTEM: BOOKINGS	DATE 25/3/87	AUTHOR: DG	Del as Required
			LDS TNF CDB

Entity ID & Name: SCHEDULED COURSE   Volumetrics: Ave 30 Max 50

Entity Description:

Key P/F	Data Items	Format	Length	Comments
✓	(COURSE CODE	X	5	} COMPOSITE KEY
	(START DATE)	X	2/3	(YY/DDD)
	END DATE	X	8	YY/MM/DD
	MAX. NO. DELEGATES	9	2	
	NO. PLACES BOOKED	9	2	
	Total Size		22	

SSADM Version 3 Issue 1   Jan 1986

Entity Description / Optimised TNF Relations

SSADM DOC: 8

Page 3 of 14

| SYSTEM: BOOKINGS | DATE 25/3/87 | AUTHOR: DG | Del as Required |
| | | | LDS NEW OLD |

| Entity ID & Name: LOCATION | Volumetrics: Ave 10 Max 20 |

Entity Description:

Key P/F	Data Items	For mat	Length	Comments
✓	LOCATION CODE	X	5	
	LOCATION NAME	X	30	
	CATEGORY	X	5	(HOTEL, COMPUTRAIN)
	ROOM	X	10	ROOM NAME
	DATES AVAILABLE ⎫	X	5	WEEK NO./DAY
	CAPACITY (REPEATING)	9	2	(E.G. 40/TU) REPEATED
	FACILITIES ⎭	X	20	

| | Total Size | 77 | |

Entity Description /
Optimised TNF Relations

SSADM DOC: 8

Page 4 of 14

SYSTEM: BOOKINGS	DATE 25/ 3 /87	AUTHOR: DG	Del as Required LDS ~~LPM~~ ~~OLDD~~

Entity ID & Name: SCHEDULED SITE	Volumetrics: Ave 10 Max 20

Entity Description:

Key P/F	Data Items	For mat	Length	Comments
✓	(COURSE CODE (START DATE)	X X	5 2/3	(YY/DDD)
	LOCATION CODE	X	5	
	Total Size		15	

SSADM Version 3 Issue 1

Jan 1986

Entity Description /
Optimised TNF Relations

SSADM DOC: 8

Page 5 of 14

SYSTEM: BOOKINGS	DATE 25/ 3 /87	AUTHOR: DG	Del as Required

LDS TNF CLDD

Entity ID & Name: JUNCTION DATA	Volumetrics: Ave 10 Max 20

Entity Description: LINK BETWEEN LOCATION AND COURSE

Key P/F	Data Items	For mat	Length	Comments
✓	LOCATION CODE	X	5	
✓	COURSE CODE	X	5	

	Total Size	10	

SSADM Version 3 Issue 1

Jan 1986

Entity Description /
Optimised TNF Relations

SSADM DOC: **8**

Page 6 of 14

SYSTEM: BOOKINGS	DATE 25/ 3 /87	AUTHOR: DG	Del as Required LDS ☑TNF☑ CLDD

Entity ID & Name: TOPIC	Volumetrics: Ave 50 Max 75

Entity Description:

Key P/F	Data Items	For mat	Length	Comments
✓	TOPIC NO.	X	5	
	TOPIC NAME	X	30	
	TOPIC CONTENT	X	120	(MAIN HEADINGS)
✱	COURSE CODE	X	5	
✱	TUTOR CODE	X	5	
	Total Size		165	

SSADM Version 3 Issue 1   Jan 1986

Entity Description / Optimised TNF Relations

SSADM DOC: 8

Page 7 of 14

SYSTEM: BOOKINGS	DATE 25/3/87	AUTHOR: DG	Del as Required
			LDS \ TNF \ CLDD

Entity ID & Name: JUNCTION DATA	Volumetrics: Ave 20 Max 40

Entity Description: LINK BETWEEN COURSE AND MACHINE TYPE

Key P/F	Data Items	Format	Length	Comments
✓	COURSE CODE	X	5	
✓	MACHINE TYPE CODE	X	3	
	Total Size		8	

SSADM Version 3 Issue 1

Jan 1986

Entity Description /
Optimised TNF Relations

SSADM DOC: 8

Page 8 of 14

SYSTEM: BOOKINGS	DATE 25/3/87	AUTHOR: DG	Del as Required LDS TNA LDD

Entity ID & Name: MACHINE TYPE | Volumetrics: Ave 6   Max 12

Entity Description:

Key P/F	Data Items	For mat	Length	Comments
✓	MACHINE TYPE CODE	X	3	
	MAKE	X	10	
	DESCRIPTION	X	20	
	Total Size		33	

Entity Description /
Optimised TNF Relations

SSADM DOC: 8

Page 9 of 14

SYSTEM: BOOKINGS	DATE 25/ 3 /87	AUTHOR: DG	Del as Required
			LDS \ TNF \ \ CLDD \

Entity ID & Name: TUTOR	Volumetrics: Ave 20 Max 30

Entity Description:

Key P/F	Data Items	For mat	Length	Comments
✓	TUTOR CODE	X	5	
	NAME	X	20	
	ADDRESS	X	120	
	TEL. NO. (HOME)	X	20	
*	TOPIC NO. (SPECIALIST. 1)	X	5	
*	TOPIC NO. (SPECIALIST. 2)	X	5	
		Total Size	175	

Entity Description /
Optimised TNF Relations

SSADM DOC: **8**

Page 1O of 14

SYSTEM: BOOKINGS	DATE 25/ 3 /87	AUTHOR: DG	Del as Required LDS \ TNF \ CDB \

Entity ID & Name: MACHINE	Volumetrics: Ave 2O Max 4O

Entity Description:

Key P/F	Data Items	For mat	Length	Comments
✓	MACHINE CODE	X	5	
	SERIAL NUMBER	X	15	
	MEMORY SIZE	X	5	
	DISK CONFIGURATION	X	10	
	COMMUNICATIONS	X	5	(SYNC, ASYNC, NIL)
	FACILITIES	X	10	(GRAPHICS, COLOUR, MONO)
*	LOCATION CODE	X	5	HOME LOCATION
	STATUS	X	2	(ON COURSE, DEVT, REPAIR)
	Total Size		57	

Entity Description /
Optimised TNF Relations

SYSTEM: BOOKINGS	DATE 25/ 3 /87	AUTHOR: DG	Del as Required
			LDS \TNF\ \CLDD

Entity ID & Name: CUSTOMER	Volumetrics: Ave 5000 Max 10000

Entity Description:

Key P/F	Data Items	Format	Length	Comments
✓	CUSTOMER NO.	X	5	
	CUSTOMER NAME	X	30	
	CUSTOMER ADDRESS (HQ)	X	120	
	CUSTOMER ADDRESS (A/C)	X	120	
	CONTACT NAME	X	30	
	CONTACT TEL. NO.	X	20	
	CATEGORY CODE	X	2	(INTERNAL, EXTERNAL, PROFORMA, ETC.)
	Total Size		327	

SSADM Version 3 Issue 1

Jan 1986

Entity Description /
Optimised TNF Relations

SSADM DOC: **8**

Page 12 of 14

SYSTEM: BOOKINGS	DATE 25/ 3 /87	AUTHOR: DG	Del as Required

LDS TNF CDD

Entity ID & Name: SCHEDULED RESOURCE	Volumetrics: Ave 30 Max 40

Entity Description:

Key P/F	Data Items	For mat	Length	Comments
✓	(COURSE CODE START DATE)	X X	5 2/3	(YY/DDD)
✓	RESOURCE CODE	X	5	
	Total Size		15	

SSADM Version 3 Issue 1

Jan 1986

Entity Description /
Optimised TNF Relations

SSADM DOC: **8**

Page 13   of   14

SYSTEM: BOOKINGS	DATE 25/ 3 /87	AUTHOR: DG	Del as Required		
			LDS	TNF	CLDD

Entity ID & Name: BOOKING	Volumetrics: Ave 250 Max 300

Entity Description:

Key P/F	Data Items	For mat	Length	Comments
✓	(COURSE CODE START DATE)	X X	5 2/3	(YY/DDD)
✓	BOOKING NO.	X	5	
	BOOKING DATE	X	8	(YY/MM/DD)
	DELEGATE NAME	X	20	
	DELEGATE POSITION	X	20	
∗	CUSTOMER NO.	X	5	
∗	INVOICE NO.	X	15	
	BOOKING STATUS	X	2	(PROVISIONAL, CONFIRMED, CANCELLED, INVOICED)

Total Size	85

Jan 1986

Entity Description /
Optimised TNF Relations

SSADM DOC: **8**

Page 14 of 14

SYSTEM: BOOKINGS	DATE 25/ 3 /87	AUTHOR: DG	Del as Required		
			LDS	TNF	CLDD

Entity ID & Name: INVOICE	Volumetrics: Ave 200 Max 1000

Entity Description:

Key P/F	Data Items	For mat	Length	Comments
✓	INVOICE NUMBER	X	15	
	INVOICE DATE	X	8	(YY/MM/DD)
*	BOOKING NO.	X	5	
	BOOKING DATE	X	8	(YY/MM/DD)
	SCHEDULED COURSE DATE	X	8	(YY/MM/DD)
	PRICE PER DELEGATE	9	5/2	(£ZZZZ9.99)
	NO. OF DELEGATES	9	2	
	(INVOICE TOTAL)			PRINTED ONLY
		Total Size	52	

SSADM Version 3 Issue 1

Jan 1986

Data Store / Entity X-Ref

SSADM DOC: **3**

Page *1* of *2*

SYSTEM: BOOKINGS	DATE 25/ 3 /87	AUTHOR: DG

Data Store Ref.	Name	LDS Entities & Structure
D1	COURSE	
D2	SCHEDULED COURSE	
D3	COMPLETION DETAILS	

Jan 1986

Data Store / Entity X-Ref

SSADM DOC: 3
Page 2 of 2

SYSTEM: BOOKINGS	DATE 25/ 3 /87	AUTHOR: DG

Data Store Ref.	Name	LDS Entities & Structure
D4	CUSTOMER	CUSTOMER R9
D5	LOCATION	LOCATION R2
D6	RESOURCE (a) TUTOR	TUTOR R4 / TOPIC SKILL R3
	(b) MACHINE	MACHINE TYPE R12 / MACHINE R13

SSADM Version 3 Issue 1   Jan 1986

16.4 The required COMPUTRAIN system

Once the current system has been defined, the required system is then produced in the second stage of SSADM. This covers a number of detailed steps and produces a fair amount of documentation. To reduce the documentation which needs to be presented in this book, the complete required system is not shown, only representative examples.

The output from this stage includes the following (descriptions given where necessary):

(a) *Level 1 DFD of the required system.*

(b) *Level 2 DFDs of the required system.*

(c) *Elementary function descriptions*: these define in more detail the functions of the lowest level DFD process boxes.

(d) *Required logical data structure.*

(e) *Entity descriptions*: a detailed definition of the attributes (fields) appertaining to each entity, together with details of size and format.

(f) *Data store/entity cross reference.*

(g) *Problem/requirements list*: in this stage entries are made in the solutions column.

(h) *Entity/event matrix*: a method of showing which events from the DFDs affect which entities on the logical data structure.

(i) *Entity life histories*: these show how entities are created, amended and deleted and are derived from the previous matrix. These help to define which events the system must handle, and the associated handling logic. They also identify those events which will not be handled by the system.

(j) *Event catalogue*: events are grouped into common function areas and appropriate details are added. In the design stages of SSADM the event catalogue is used to define the logical process outlines.

(k) *I/O description*: a form enabling the analyst to define all the input and output for the system. At the design stage these will then be translated into the appropriate I/O layouts (see screen layouts on pp. 341–342).

To complete the analysis phase the following would be added:

(l) *The technical option details*. that is, details of the hardware and software to implement the required system.

(m) *A cost/benefit analysis* of the final solution selected.

(n) *An implementation plan and timescale* for the total project.

Having produced the final specification of the option selected, stages 4 and 5 of SSADM are then completed to define the data and processes in more detail, and stage 6 incorporates the implementation and training tasks. Since the analysis has been done to a very low level of detail, the design stages of SSADM become an extension of these preceding stages. The emphasis on very detailed analysis before design is an important strength of SSADM.

Problem / Requirements List

SSADM DOC: 2

Page 1 of 5

SYSTEM:	BOOKINGS				DATE 25/3/87	AUTHOR: DG	
		Problem Requirements			**Solutions**		
P/R Ref No.	User Name/ Ref	Prio-rity	Brief Description	Ext Nar Ref	DFD/ LDS Ref	Brief Description	Ext Nar Ref	DFD/ LDS Ref
P001	BOOK	1	Increased volumes of bookings incurring large amounts of paperwork and manual look-up and cross-referencing.		1.3,1.2/ 1.5/ C5/C6	On-line input of bookings, on-line enquiry, automatic cross-referencing.		1.1,1.2, 1.3,1.4, 1.6
P002	BOOK	1	Manual preparation of joining instructions for all courses.		1.4/ C8	Automated print of joining instructions 3 weeks prior to course start.		1.5
P003	BOOK	1	Any necessary correction of errors is very time-consuming because of manual matching and cross-referencing method.			On-line input and automatic cross-referencing will remove this.		ALL
P004	BOOK	1	The overall workload leaves less and less time for the prime priority – satisfying the customer.			Automation will reduce clerical workload substantially.		ALL

SSADM Version 3 Issue 1

Jan 1986

Problem / Requirements List

SSADM DOC: 2

Page 2 of 5

SYSTEM:	BOOKINGS				DATE 25/3/87	AUTHOR: DG

		Problem Requirements			Solutions		
P/R Ref No.	User Name/ Ref	Prio-rity	Brief Description	Ext Nar Ref	DFD/ LDS Ref	Brief Description	Ext Nar Ref	DFD/ LDS Ref

P/R Ref No.	User Name/Ref	Prio-rity	Brief Description	Ext Nar Ref	DFD/LDS Ref	Brief Description (Solutions)	Ext Nar Ref	DFD/LDS Ref
R001	BOOK	1	Up-to-date database of customers and courses to be held at Computrain. This to supercede parent company listings.			This to be provided as the basis of the new system together with details of resources (i.e. tutors and machines.)		4
R002	BOOK	1	On-line enquiry and update facilities for provisional and confirmed bookings together with amendments and cancellations.			To be provided as for R001 above.		1.1,1.2, 1.3,1.4, 1.0
R003	BOOK	1	Automatic print of joining instructions for course delegates 3 weeks prior to the start of the course.			To be provided as for R002 above.		1.5

SSADM Version 3 Issue 1

Jan 1986

Problem / Requirements List

SYSTEM: BOOKINGS

SSADM DOC: 2

Page 3 of 5

DATE 25/3/87 AUTHOR: DG

P/R Ref No.	User Name/ Ref	Prio-rity	Problem Requirements			Solutions		
			Brief Description	Ext Nar Ref	DFD/ LDS Ref	Brief Description	Ext Nar Ref	DFD/ LDS Ref
R804	BOOK	2	Monthly production (automatic) of management reports and associated batch enquiry facilities.			Automatic printing of monthly management reports to be provided. Enquiry facilities will be a future enhancement.		5
P805	INV	1	Delay in update by parent company of customers and courses. This results in duplicate set-up forms being completed.			The provision of a local Computrain database of customers and courses will remove this.		4
P806	INV	1	Unnecessary duplication of effort in the preparation of invoices. Manual preparation at Computrain, dummy-run of input on mainframe, check against originals, resubmission of corrections and final invoice print from mainframe.			At the end of each month, invoice/credit note details will be printed at Computrain and sent to the parent company together with 'floppy disk' for mainframe input of data.		3

Jan 1986

Problem / Requirements List

SSADM DOC: 2

SYSTEM: BOOKINGS							DATE 25/3/87 AUTHOR: DG		
		Problem Requirements				**Solutions**			
P/R Ref No.	User Name/ Ref	Prio- rity	Brief Description	Ext Nar Ref	DFD/ LDS Ref	Brief Description	Ext Nar Ref	DFD/ LDS Ref	
P007	INV	1	Errors created by pressure of work during the month, and these compounded by the monthly invoice deadline.			Resolved as for P005 above.		3	
R005	INV	1	Automatic production of customer and course lists on an ad-hoc basis from local Computrain database.		/C1, C2	To be provided as for P005 above, with selective extract and print facilities.		4	
R006	INV	1	Automatic monthly production of billing/credit note requests for sending to the parent company.			To be provided as for P006 above.		3	

SSADM Version 3 Issue 1

Jan 1986

Problem / Requirements List

SSADM DOC: 2

Page 5 of 5

SYSTEM: BOOKINGS						DATE 25/ 3 /87	AUTHOR: DG		
		Problem Requirements				**Solutions**			
P/R Ref No.	User Name/ Ref	Prio-rity	Brief Description	Ext Nar Ref	DFD/ LDS Ref	Brief Description	Ext Nar Ref	DFD/ LDS Ref
R007	INV	1	Input of all resource availability (tutors and machine) and automatic scheduling facilities together with exception reporting and on-line scheduling facilities.			On-line update of relevant resource data stores to be provided, together with batch scheduling and on-line rescheduling facilities.		2
R008	MGMT	1	Production of monthly statistics e.g. (a) Revenue : actual v forecast (b) Revenue by tutor (c) Revenue by course type			To be provided from course history information (e.g. completion details)		5

SSADM Version 3 Issue 1

Jan 1986

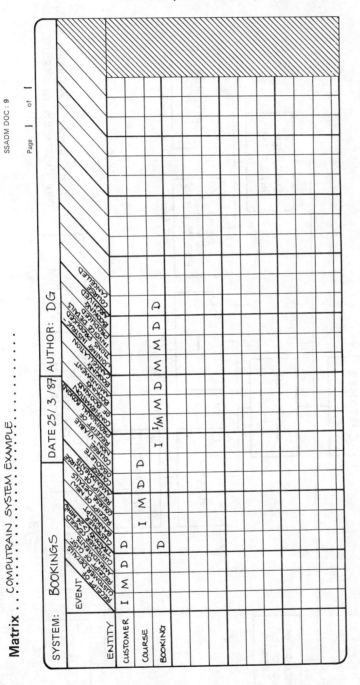

Matrix . . COMPUTRAIN SYSTEM EXAMPLE. . . . .

SSADM DOC : 9

Page ⎯ of ⎯

SYSTEM: BOOKINGS   DATE 25/3/87   AUTHOR: DG

Jan 1986

SSADM Version 3 Issue 1

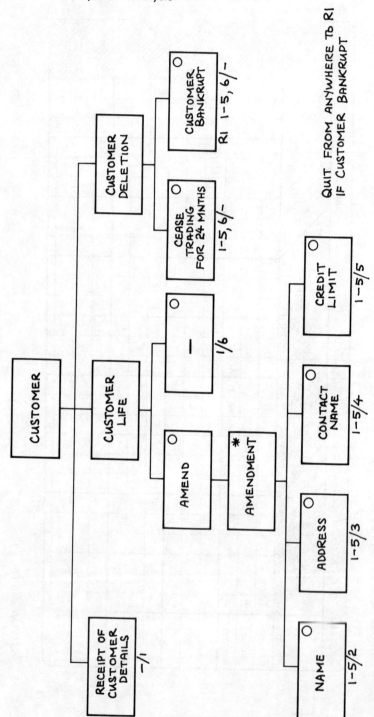

Fig. 16.4.1 ELH – customer

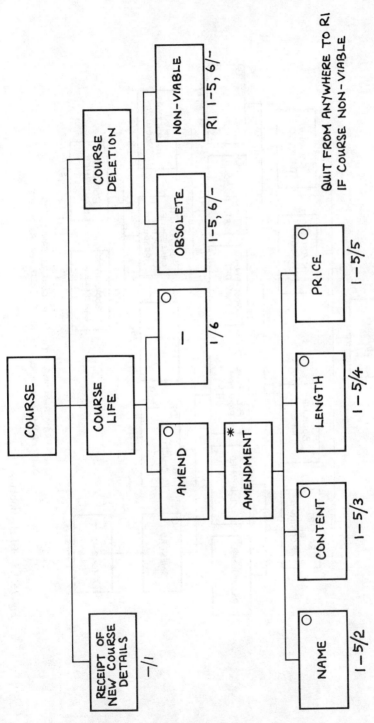

Fig. 16.4.2 ELH – course

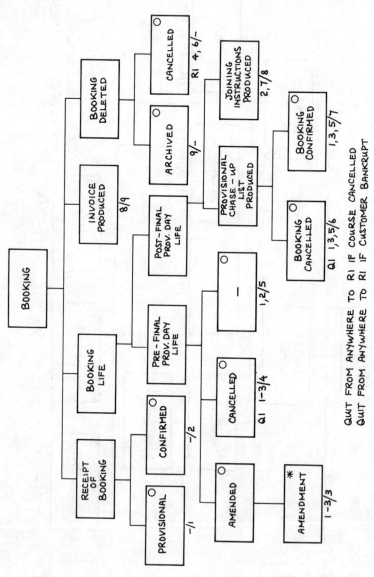

Fig. 16.4.3 ELH – booking

Event Catalogue

SYSTEM: BOOKINGS	DATE 25/ 3 /87	AUTHOR: DG

SSADM DOC: 11

Page 1 of 2

Process / Function ID:

Process / Function Name: BOOKING DETAILS

Event ID	Event Name	Event Description	Entities Affected	Effect Type (I/M/D)	Per	Volumetrics Max	Ave	Comments
001	PROVISIONAL BOOKING	The receipt of a provisional booking for a scheduled course.	BOOKING SCHEDULED COURSE	I M	MTH	300	250	
002	CONFIRMED BOOKING	The receipt of a confirmed booking, or the receipt of a confirmation of an existing provisional booking.	BOOKING SCHEDULED COURSE	I / M M	MTH	300	250	
003	BOOKING AMENDMENT	A provisional or confirmed booking is amended by the customer (ie. name of delegate.)	BOOKING	M	MTH	20	5	
004	BOOKING CANCELLATION	A provisional or confirmed booking is cancelled by the customer.	BOOKING SCHEDULED COURSE	D M	MTH	30	15	

SSADM Version 3 Issue 1

Jan 1986

Event Catalogue

SSADM DOC: 11

Page 2 of 2

SYSTEM: BOOKINGS	DATE 25/3/87	AUTHOR: DG

Process / Function ID:

Process / Function Name: CUSTOMER DETAILS

Event ID	Event Name	Event Description	Entities Affected	Effect Type (I/M/D)	Volumetrics			
					Per	Max	Ave	Comments
005	RECEIPT OF NEW CUSTOMER DETAILS	A new customer supplies his details. This could happen as part of a course enquiry or a booking.	CUSTOMER	I	MTH	30	20	
006	RECEIPT OF CUSTOMER AMENDMENT	A customer sends in an amendment to his current details (e.g. name, address, contact.)	CUSTOMER	M	MTH	50	20	

SSADM Version 3 Issue 1

Jan 1986

I/O Description

SSADM DOC: 7
Page 1 of 2

SYSTEM: BOOKINGS	DATE 25/ 3 /87 AUTHOR: DG

Type : Screen / ~~Print~~ / ~~Document~~

~~I/O Ref No.~~ / I/O Name / ~~Flow Name.~~   BOOKING DETAILS

Description   In order to update / amend / delete bookings on a
scheduled course, the course details are displayed on the
screen and the relevant booking details are entered, changed
or deleted.

Data Items	Format	Length	Comments
Todays date	X	8	(DD/MM/YY)
Course code	X	5	
Start date	X	8	(DD/MM/YY)
Course name	X	30	
Maximum no. of places	9	2	
BOOKING DETAILS			Maximum of 9 on first screen and 16 on each continuation screen up to a theoretical maximum of 99
Booking number	X	5	
Booking date	X	8	(DD/MM/YY)
Booking status	X	2	* See entries below
Delegate name	X	20	
Delegate position	X	20	
Customer number	X	5	
			* Booking status
			PR = Provisional
			CF = Confirmed
			JI = Joining Instructions produced
			AM = Amended
			IN = Invoiced
			CA = Cancelled

I/O Description

SYSTEM: BOOKINGS	DATE 25/ 3 /87	AUTHOR: DG

Type : Screen / ~~Print~~ / ~~Document~~

~~I/O Ref No.~~ / I/O Name / ~~Flow Name.~~ CUSTOMER DETAILS

Description Using this screen details may be entered for a new customer or amendments entered for an existing customer. The customer CANNOT be deleted using this screen.

Data Items	Format	Length	Comments
Todays date	X	8	(DD/MM/YY)
Customer number	X	5	
Category	X	2	External, internal, proforma
Customer name	X	30	
Address (HQ)	X	120	} 4 address lines x 30
Address (A/C)	X	120	
Contact telephone no.	X	16	
Contact name	X	30	

SCREEN LAYOUT

SYSTEM __COMPUTRAIN__ SUB-SYSTEM __BOOKINGS/INVOICING__ DATE __25/3/87__

FUNCTION NOs. __1.1, 1.2, 1.3, 1.4__

TRANS ID's __PROVBOOK, CONFBOOK, AMNDBOOK, CANCBOOK__   SCREEN No. __1__

```
 1234567891012345678920123456789301234567894012345678950123456789601234567897012345678980

      COMPUTRAIN BOOKING SYSTEM                    DATE 25/03/87

      COURSE CODE - ICM02   MAX PLACES = 12   START DATE - 17/04/87

      COURSE NAME - INTRO. TO STRUCTURED ANALYSIS

      BOOKING DETAILS
      BOOKING  BOOKING   DELEGATE      DELEGATE      CUSTOMER
      NO.      DATE      STATUS  NAME  POSITION      NUMBER

      ISA01   16/02/87   CD   D.HANSEN        JR. PROGR     A0014
      ISA05   17/02/87   PR   A.LONGFELLOW    TR. ANALYST   C0093
      ISA06   14/01/87   CD   P.THOMPSON      AN. PROGR     H0049
      ISA10   16/03/87   PR   H.ARBUTHNOT     AUDITOR       P1672




      BOOKING:  PR=PROVISIONAL, CF=CONFIRMED, CA=CANCELLED, AM=AMENDED
      UPDATE - ENTER DETAILS + CTRL/ENTER
      EXIT - CTRL/Q
```

SCREEN LAYOUT

SYSTEM ___COMPUTRAIN___ SUB-SYSTEM ___BOOKINGS/ INVOICING___ DATE __25/3/87__

FUNCTION NOs. __4__

TRANS ID's __ADDCUST, AMDCUST__ SCREEN No. __2__

```
COMPUTRAIN BOOKING SYSTEM                    DATE 25/Ø3/87

          CUSTOMER DETAILS
          (UPDATE/AMEND)

CUSTOMER NO.      - HØØ47      - CATEGORY - EX

CUSTOMER NAME     - HANSONBY SOFTWARE AND PARTNERS

ADDRESS (H/Q)     - 14, PETTERIDGE LANE,
                    ROCHESTER
                    KENT ME2 4XY

ADDRESS (A/C)     - 27, FORTESCUE DRIVE,
                    HADLOW-ON-SEA
                    ESSEX HA6 5NE

CONTACT TEL NO.   - 086-32-5187

                                  DAVID HUNGERFORD-SMITH



UPDATE - ENTER DETAILS + CTRL/ENTER
EXIT   - CTRL/Q
```

17 Financial Applications

Systems analysts are likely to be asked to study, or to become involved in, the implementation of various sorts of accounting systems. It is therefore important to understand the workings and the objectives of these systems. There are two aspects of the accounting function which need to be considered: financial accounting and management accounting.

In medium-to-large companies the functions of financial accounting and management accounting will usually be performed by separate sections, but in smaller organizations they are often combined. The role of the financial accountant is to record the day-to-day transactions of the enterprise and at the end of the accounting period to prepare a summary statement of performance, called a profit and loss account, and to produce a snapshot of the company's financial standing, called a balance sheet. These statements are prepared primarily for publication to shareholders, to the public at the registry, called Companies House, and for submission to the Inland Revenue in order to calculate the liability for corporation tax at the end of the year.

However, these financial statements are of little use in the management of the business since they are prepared too infrequently. The management accountant, therefore, has the much more immediate role of providing the information necessary to ensure the business remains on course with its objectives.

17.1 The business cycle

To understand the role of the accounting function it is necessary to understand the flow of value which the accountant is endeavouring to report. Businesses have a flow of value in two directions. There is an inflow of value and an outflow of value. If the inflow of value exceeds the outflow of value the business will survive and prosper, but if the outflow of value exceeds the inflow, then the business will

344 Basic Systems Analysis

be able to sustain itself only for a short while by living off its past reserves and will eventually be forced to close down. Value coming into a business and value going out of a business arise in various ways and flow right the way through the business, being transferred as they go. This transfer cycle in a typical manufacturing business can be illustrated in the following way.

To start a business the proprietor will need some capital which will be received into the business in the form of cash. This cash will be used to purchase assets such as plant and machinery or furniture; because of their more permanent nature these are described as fixed assets. More of the cash will be used to purchase stocks and raw materials which will subsequently be turned into work-in-progress and with further work become finished goods. All of these items are an outflow on the business.

Inflow to the business arises from the sale of manufactured goods and if, as they should be, they are sold at a greater value than the cost of production and sales combined, the profits arising will also represent an inflow of value. There are in addition two other ways in which value flows in or out of the business. When raw materials are required it is usual for the supplier to grant a period of credit before payment is due. This 'borrowing' therefore represents an inflow to the business just as a loan from a bank would represent an inflow. Similarly, the sale of the manufactured product will be undertaken on terms under which the purchaser will not be expected to pay for the goods immediately on taking possession of them. Therefore the granting of this credit period represents an outflow for the business, which will only be made good when the cash is received. These inflows and outflows are illustrated in Figure 17.1.1 (see next page).

17.2 Financial accounting

The role of the financial accountant is to record the constant movements of value in and out of the business. This is done by using daybooks, ledgers and journals, and from these documents the transactions will subsequently be summarized into a general or nominal ledger, from which will be prepared the balance sheet and profit and loss account.

It is essential that an up-to-date and accurate record is kept of the value expended and the value received by the business, and a constant record of the amounts owned by the business to suppliers

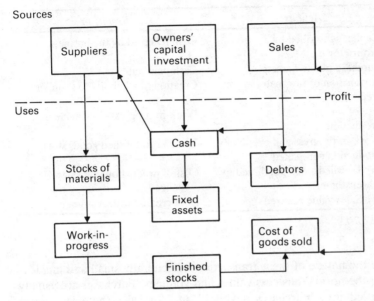

Fig. 17.1.1

and the amounts it is owed by customers. Each of these transactions, whether buying or selling, represents a flow of value, and causes two accounting entries to be made. When raw material is purchased, raw material stock is increased, but there is an equal flow of value in cash out of the business or the creation of an equivalent liability to the supplier if the raw materials are purchased on credit. Similarly, when the business sells goods, either cash is received or a debt is created with that customer. When the customer eventually settles his debt, two further transactions take place: firstly the cash is received and secondly the debt owed by the customer is eliminated. This can be illustrated in Figure 17.2.1.

Although it would be possible to record all the transactions and their values, as is shown in Figure 17.2.1, it would not be a very practical method, since they could not readily be summarized, nor could accounts be prepared easily. Nevertheless, the principle is evident of the two-sided nature of each transaction and a financial accountant adopts a similar approach in the form of double entry book-keeping. For this a set of books is needed to record each type of transaction. There is therefore a book to record cheque and credit transactions called the 'cash book' and a book to record transactions actually undertaken in cash called the 'petty cash book'

Actions	Results in
1. Capital subscribed by a proprietor	Cash received by the business
2. Purchase of fixed assets	Cash paid out
3. Acquisition of raw materials on credit	Creation of a liability to supplier
4. Use of raw materials in production	Work-in-progress — asset created
5. Work-in-progress and eventually production completed	Transfer to finished goods stock
6. (i) Finished stock purchased by customer	Cost of production and sales realized
(ii) Sales value realized	Debt created with customer

Fig. 17.2.1

since the nature of these transactions is generally small and incidental to the normal business values and volumes. Purchases are usually recorded in a 'purchases daybook' and for sales there is a 'sales daybook'. Credit liabilities with suppliers and the debtors for sales are entered in 'purchase' and 'sales' ledgers respectively. To record each transaction the convention of debits and credits is adopted. A series of typical transactions in the books of account is shown in Figure 17.2.2, in which the following transactions have taken place:

(1) The proprietor submits £4000 in cash to start his business and this is paid into the bank.

(2) The business purchases £1500 of fixed assets and these are paid for by cheque.

(3) Raw materials are purchased on credit from company A Limited to the value of £2000.

(4) Wages amounting to £500 are paid in cash, which is drawn from the bank.

(5) Sales of £500 are made to company M Limited on normal credit terms.

(6) The business settles its liability to company A Limited for the raw materials purchased by cheque.

(7) A cheque is received from company M Limited to pay for the goods sold to them.

The debit entries and the credit entries for each transaction are recorded firstly in journals and then shown in the books of account in the manner shown in Figure 17.2.3.

Debit	£	Credit	£
1. Cash book	4 000	Proprietors' subscription account	4 000
2. Fixed assets	1 500	Cash book	1 500
3. Raw materials purchased	2 000	Purchase ledger account with A Limited	2 000
4. Wages account	500	Cash book	500
5. Sales ledger account with M Limited	500	Sales	500
6. Purchase ledger account with A Limited	2 000	Cash book	2 000
7. Cash book	500	Sales ledger account with M Limited	500

Fig. 17.2.2

By recording the entries in this way we have not only analyzed and summarized the transactions of the business, but we have constructed the books in such a way that they can be checked for arithmetical accuracy through a trial balance.

17.2.1 Trial balance

Because all of the transactions have both a debit entry and a credit entry, it follows that a summary of the entries will comprise debit entries and credit entries of equal value. Consequently, it can be proved that the books have been maintained accurately by testing them through the production of a trial balance. This is done by identifying the balance of the entries of each account and by scheduling them on a summary. A trial balance for the simple limited transactions shown in Figure 17.2.2 has been produced and is shown as Figure 17.2.4. A full set of books would of course yield many more entries for this trial balance.

While double entry book-keeping and a trial balance can help to ensure that books of account have been maintained accurately, it is not possible to say with absolute certainty that the books are correct. It would be possible, for example, for the trial balance to agree even if a complete transaction had been omitted, since both the debit and credit entries would be absent from the records.

Cash book

	£		£
1. Received from proprietor	4 000	2. Paid for fixed assets	1 500
7. Received from M Limited	500	4. Paid for wages	500
		6. Paid to A Limited	2 000

Proprietors account

	£		£
		1. Money received	4 000

Fixed asset account

	£		£
2. Fixed assets purchased	1 500		

Raw materials

	£		£
7. Acquired for A Limited	2 000		

A Limited

	£		£
6. Paid to A Limited	6 000	3. Raw materials purchased	2 000

Wages

	£		£
4. Wages paid	500		

Sales

	£		£
		5. Sales to M Limited	500

M Limited

	£		£
5. Sales supplied on credit	500	7. Received from M Limited	500

Fig. 17.2.3

Equally, the books would be incorrect if an entry, say for £200, was incorrectly recorded both as a debit and a credit for £300. The trial balance would still agree.

However, from the trial balance and a valuation of stock and work-in-progress the financial accountant can draw up his profit and loss account and his balance sheet. Examples of these are shown in Figures 17.2.5 and 17.2.6.

Trial Balance as at 31 December 1984

	Debit £	Credit £
Cash balance	500	
Proprietors' account		4 000
Fixed assets	1 500	
Raw materials	2 000	
Credit account with A Limited	Nil	Nil
Wages	500	
Sales		500
Debtors' account with M Limited	Nil	Nil
	£4 500	£4 500

Fig. 17.2.4

XYZ Ltd
Profit and Loss Account for the year ended 31 December 1984

	£	£
Sales		150 000
Cost of raw materials:		
Opening stock	25 000	
Purchases	110 000	
	135 000	
Less closing stock	35 000	
	100 000	
Wages and salaries	20 000	
Rent and rates	10 000	
Light and heat	8 000	
Travelling and entertainment	5 000	
		143 000
Net profit for year		7 000
Cumulative profit brought forward		17 000
Cumulative profit carried forward		24 000

Fig. 17.2.5

Balance Sheet
as at 31 December 1984

	£	£
Fixed assets		30 000
Stocks	35 000	
Debtors	20 000	
Cash at bank	4 000	
	59 000	
Creditors	35 000	
		14 000
		44 000
Represented by share capital		20 000
Accumulated reserves		24 000
		£44 000

Fig. 17.2.6

17.3 Management accounting

The financial accounts are the basis of other management information since they are in themselves inadequate for management control purposes for a number of reasons.

17.3.1 Frequency

Financial accounts are prepared probably only once or perhaps twice a year and therefore cannot give any immediate information on the position of the company. To control and monitor the company's performance, information needs to be provided more regularly, and this is usually done on a monthly basis.

17.3.2 Analysis

While the financial accounts demonstrate the totals of revenues and costs incurred on, for example, wages, rents of premises, depreciation of fixed assets etc., they cannot answer the sort of question that management needs to have answered if it is to manage the business, such as:

(1) Which product within the range is the most profitable?

(2) Is it more cost effective to make product A ourselves or to have it manufactured by a sub-contractor?

(3) What are the sales and marketing costs of the business?

and many others.

17.3.3 Comparison with objectives

It is likely that the financial accounts will demonstrate the company's performance compared with the previous financial year. This measure is unlikely to be adequate as it provides no measure against the company's current objectives. These objectives are usually assessed in the form of budgets established at the beginning of a financial year and we shall discuss this in more detail later in Section 17.5.

To deal with these types of deficiencies the management accountant must take the financial information and analyze it in a different way, so whereas the financial accounts show the information as illustrated in Section 17.2 the management accounts will probably show it in different ways and on a more frequent basis. It is likely for example that costs will be analyzed for each product produced. Such an analysis might look as shown in Figure 17.3.1. From this analysis we can see that Product C is not as profitable as Product A or B. Alternatively, accounts can be prepared to show costs analyzed by function (Figure 17.3.2).

The management accountant therefore tries to identify the incidence of costs in the various sections of the business and to establish cost centres, into which he can analyze his overall financial costs. These cost centres could relate either to the various sections of the business or to jobs products or services within the organization. It must be stressed that the management accountant should do the analysis solely on the basis of a thorough understanding of the operation of the organization and not have pre-determined ideas about how the analysis should be made.

Quite often a cost centre is simply a department or section in which related operations are performed. There is no general rule about what constitutes a cost centre and, in fact, any physical location, piece of work, person or group of people, process or part of a process can be classified as a cost centre to which data will be specifically channelled. In practice, the main cost centres will be divisions of the firm's functional areas: production, selling, distribu-

	Total £	Product A £	Product B £	Product C £
Sales	150 000	60 000	40 000	50 000
Direct costs:				
Raw materials	100 000	40 000	20 000	40 000
Direct wages	15 000	6 000	7 000	2 000
Rent and rates	7 500	2 500	2 500	2 500
Light and heat	6 000	2 000	2 000	2 000
	21 500	9 500	8 500	3 500
Overheads:				
Wages and salaries	5 000			
Rent and rates	2 500			
Light and heat	2 000			
Travelling and entertainment	5 000			
Net profit	£7 000			

Fig. 17.3.1

	Total	Production	Sales	Admin
Raw materials	100 000	100 000		
Wages and salaries	20 000	15 000	2 000	3 000
Rent and rates	10 000	75 000	1 600	900
Light and heat	8 000	6 000	1 200	800
Travel and entertainment	5 000		4 000	1 000
	143 000	128 500	8 800	5 700
Sales	150 000			
Net profit	£7 000			

Fig. 17.3.2

tion, administration, research. These can be subdivided. For instance, the production function as a whole could become a number of production cost centres where work on the saleable products is carried out (e.g. machine shop, assembly shop), together with a number of service cost centres where the products made (e.g. small tools in a toolroom) or services rendered (e.g. canteen, maintenance, stores, inspection) are not directly part of the saleable product costs, but are part of the overall cost of providing production facilities. In other functional areas, cost centre analysis may be appropriate to such locations as a typing pool, branch office, warehouses, vehicle maintenance shops, vehicle groups or laboratories.

For cost centres and responsibility centres the most appropriate size has to be chosen and it is not possible to generalize. An impersonal centre may be limited to a particular machine and a personal centre could even consist of a single person.

Therefore the information required from the management accounts will depend entirely upon the circumstances of the business itself and how it is controlled. It is quite probable that the management accountant will prepare several sets of accounts, analyzed in these different ways, for the various senior managers of the company. What should be perfectly clear is that the management accountant can perform his role only after a thorough understanding of the operation of the business. However, to analyze the financial information in a way useful to management much will depend on the reliability of the underlying data or statistical analysis upon which the management accountant's judgements will be based.

17.4 Costing

The costing function of an enterprise attempts to measure the cost of a product, of a service or an operation, so that management can judge the efficiency of the enterprise and establish the price or the cost of these products or services. It is the management accountant's role to collate and prepare the information to enable these judgements to be made.

Costs fall into two categories: those which can be identified directly with the activity that generates them, such as raw materials and labour costs in the manufacturing department. These are called direct costs. Costs which cannot be directly identified, either with products or specific sections of the business, for example, telephone costs or the costs of the accounts department or the personnel department, are called indirect costs. They are also sometimes described as overhead costs.

Direct costs are sometimes further analyzed to distinguish between those which are variable in relation to the activity of production and those costs which are fixed and don't vary as production changes. For example, the costs of the production department in terms of its rent, rates and machine costs are unlikely to vary with the output produced. The cost of raw materials, however, will vary directly with output produced.

It is very difficult to establish the precise cost of any product or service. It is probable that costs, when analyzed, do not react to changes in production or business activity in quite the same way as expected and therefore variable costs may not react quite so accurately to changes in the level of activity, as was thought at first. Some costs are shared with other parts of the business: for example, the cost of premises, which may have a production and an administrative function, may have to be analyzed on the basis of some acceptable assumption about these two functions. Further, the cost attributable to items such as plant and machinery is based on a forecast of the expected life of that plant and machinery in the business. In most cases it is impossible to forecast this with any precise degree of accuracy and therefore the proportion of the total cost of the plant charged each year to the production cost centre will, by its very nature, be only an estimate. Eventually, however, when all the elements of cost *have* been identified, it is possible for the management accountant to bring together a structure for this reporting along the lines shown in Figure 17.4.1.

The analysis of all these costs can be achieved by a reliable system of coding. The general principles of code design were described in Chapter 4 and here it suffices to summarize by saying that the objective of coding for costing purposes is firstly to identify direct material, direct labour and then the other direct expenses attributable to the products and services. Secondly, to identify production, selling and administration and any other functions within the business, and thirdly, within each function to capitalize each department or cost centre. Through this coding method the costs of the products and services of the business can be analyzed and management should be able to determine the costs applicable to each aspect of the enterprise.

There are in common use a number of recognized costing methods, of which the following five are typical:

(1) *Job or contract costing* where each job, contract or project is separately charged, such as the manufacture of a special machine or an individually designed building.

(2) *Batch costing* where jobs are organized in batches with each batch regarded as a cost unit—for example, in the manufacture of sets of components or loaves of bread.

(3) *Unit costing* here manufacture is easily measured in identical units of output, such as tons of steel or square metres of glass.

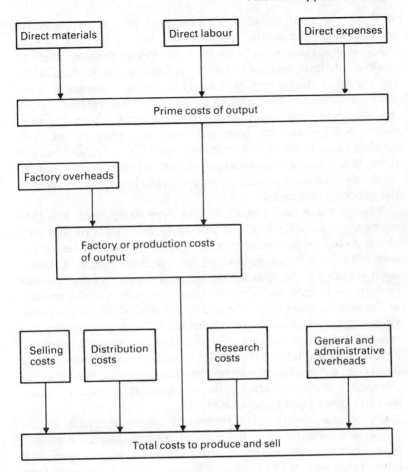

Fig. 17.4.1

(4) *Operational costing* this is similar to unit costing but applies to services, so that for example in the transport industry the ton mile becomes the unit of cost, or in hospitals, the patient's day becomes the unit of cost.

(5) *Process costing* standardized items are produced in a continuous process of production. Costs are allowed to flow for a period of time and the output is then used to determine an average cost per unit of measure—for example, in the production of cement, chemicals, or consumer durables assembled on a continuous flowline.

It should be clear therefore that the type of job or production method will substantially influence the costing method chosen. Additionally, there will always be a number of problem areas to resolve, whatever method of costing is used, and however efficient the system of coding may be it would be wrong to assume that the distribution of overhead costs to their relevant cost centres is easily achieved. Although some costs can be directly allocated, many others will be common costs and some way must be found of dividing these between their relevant cost centres. The only way of doing this is to examine each type of cost so that a decision can be made as to the most sensible method of spread or apportionment to the various cost centres.

When departmental cost totals have been determined, it is then necessary to consider how best to recover these costs by making a charge to the product being produced. Once again it is necessary to establish a basis for doing this and one common method of achieving it is to divide the total cost by the total measure of the activities performed. It must be appreciated that cost rates will absorb costs as the outputs occur but at the end of a period of time when a total valuation is made it will be found that under- or over-absorption has resulted due to cost changes or volume changes. It may therefore be necessary to use further techniques to give better control over the cost relationships. There are therefore two important costing methods used to improve control of the management of costs. These are standard costing and marginal costing.

One of the defects of a traditional costing approach is that historical information is produced and costs are accepted as they have occurred. While this information is useful and goes well beyond that provided by the financial accounts, no thought has been given to what the costs should have been. Historical costing information can assist management to carry out further cost studies so that targets can be set for the future through the setting of standards. These standards represent pre-determined cost allowances against which actual costs can be compared, so that the differences —or variances, as they are known—can be identified. These variances can then be investigated so that causes are located and remedial action taken.

Standards can be set for any item of cost, both as to price and quantity in any cost centre situation. This can be arranged either on an *ad hoc* basis or as a complete standard-costing system with reports designed to identify variances and to report them to appropriate management. A standard cost is not just an estimate in the accepted sense. It should be scientifically based, obtainable as a target, and most importantly, agreed by all parties using it. In a full standard-costing system it should also be possible to standardize all types of output and to express these in common terms. Output targets and cost targets can then be related so that variance analysis becomes meaningful, since there is a message to be learned both from adverse and from favourable variances. Similarly, targets and prices can be set for sales, so that actual performance compared with standards will give rise to sales variances.

In addition to analyzing expenses according to the nature of the expense or the section of the enterprise concerned, expenses can be classified according to the way the level of the cost incurred for each expense reacts to changes in the volume of production or activity. Three main behavioural groups of costs can be identified. There are:

(1) *Variable costs*   that tend to vary directly and proportionately with the volume of production
(2) *Fixed costs*   those that tend to be unaffected by changes in the volume of production
(3) *Semi-variable costs*   that tend to rise or fall along with changes in volume but not necessarily in direct proportion to the changes in volume.

The costs in this third group can be studied by making an analysis of the cost and volumes over a number of time periods so that the fixed elements can be separated fairly accurately.

This analysis can be further extended to the main types of production so that the product margin or contribution can be calculated. Contribution is the gap between sales and variable costs. Total contribution accruing must first recover the fixed costs and any surplus will be profit. This can be stated as an equation:

$$\text{sales} - \text{variable cost} = \text{fixed cost} + \text{profit (or loss)}$$

Break-even point then is that level of activity at which contribution exactly recovers the fixed costs so that there is neither profit nor loss. Marginal costing information is then often presented in the form of a break-even chart or graph and it is possible to use this graphical technique to forecast profit return in alternative situations. This marginal costing approach is useful for the following four purposes:

(1) Making an assessment of the product mix for the selection of those products which give best contribution in relation to volumes that can be made and sold.

(2) In the fixing of selling prices where it can be used as a short-term expedient in times of high competition, or to introduce new products.

(3) For make or buy decisions where suppliers' prices can be compared with marginal cost, provided that the capacity in-house being considered cannot be used to make an alternative contribution.

(4) *In planning* A study of the effects of volume and prices on contribution can assist in the formulation of future plans, especially where increased fixed costs are envisaged through investment decisions.

17.5 Measures of performance

From the information available to the enterprise the financial or the management accountant, the bank manager and investment analysts will apply performance measurement techniques for a variety of reasons.

The bank manager will need to know whether the finance he is being asked to provide is secure. The investment analyst will compare the performance of one company with another to determine in which company he should advise his clients to invest. The business manager and his accountant will want to know that the company's potential is being realized.

There are many methods and techniques for doing this and it is important for the systems analyst to appreciate one or two of the more common ones. It must be emphasized, however, that no single performance ratio will identify all the criteria necessary and it will be up to the judgement of the person reviewing the figures to decide the most appropriate measure of performance in the circumstances. Some of the more widely used ratios are:

(1) *Gross profit as a proportion of sales* Derived from the management accounts.

(2) *Net profit to sales* Derived from the management and financial accounts.

(3) *Yield on capital employed* From the financial accounts.

(4) *Stock turnover* Based on information in the management and financial accounts.

(5) *The proportion of debtors to sales*   Based on information in the management and financial accounts.

(6) *Gearing*   From the financial accounts.

For each of these measures of performance we give a brief description and an illustration.

17.5.1 Gross profit to sales

This is probably the most commonly used measure of performance. It is used to enable a business to identify when its profit margin or contribution is being eroded by falling sales prices or increasing overheads. This performance measure is very general and is not able to reflect changes due to an altered product mix.

Illustration
Company A markets three products—Alpha, Beta and Gamma; sales and the cost of sales last year were as follows:

	£	Cost of sales £
Alpha	100 000	50 000
Beta	120 000	90 000
Gamma	180 000	60 000

Firstly, we calculate the gross profit or margin for each product, and then that of the company overall:

	Alpha £	Beta £	Gamma £	Total company £
Sales	100 000	120 000	180 000	400 000
Cost of sales	50 000	90 000	60 000	200 000
Gross profit	50 000	30 000	120 000	200 000
Percentage margin	50%	25%	66⅔%	50%

In the current financial year, sales of products Alpha and Gamma are expected to remain the same and deliver the same margin percentage. Product Beta's sales, however, are expected to double while the margin percentage is expected to remain the same. The effect on the overall company margin can be calculated as follows:

	Alpha £	Beta £	Gamma £	Total £
Expected sales	100 000	240 000	180 000	520 000
Percentage margin	50%	25%	66⅔%	?
Actual margin	50 000	60 000	120 000	230 000
Revised overall company margin				44.2%

17.5.2 Net profit to sales

This ratio is used as an internal measure to control overheads, and externally as a comparison of the business's efficiency. The important point to remember is that businesses have a variety of trading methods and it is therefore essential that the company's accounting policies for calculating turnover are the same when comparisons are being made. Calculations are performed on the net profit before tax and also occasionally before interest on borrowing.

Illustration

Company X has a turnover of £500 000 and a net profit before tax of £40 000. What return on turnover is the company achieving?

$$\frac{\text{Profit before tax}}{\text{Turnover}} \quad \frac{40\ 000}{500\ 000} \times 100 = 8\%$$

17.5.3 Yield on capital employed

From an investment point of view it is necessary to ensure that the return being made measures up to alternative investment opportunities.

This ratio is used to determine which of two or more competing investment projects should be undertaken, to ensure the company as a whole is operating with an efficient use of funds, or in a takeover situation. In this last case the company making the bid can assess what level of premium they are prepared to pay in order to make the investment in the takeover worth while.

Illustration

Company A Limited has two alternative investment opportunities open to it. It can either set up a plant costing £500 000 to make a particular product or it can make an approach to take over an existing company which already makes that product, but where the cost is expected to be £600 000 including goodwill. It is estimated that from its own facility maintainable profits would be £75 000 per year. The existing company is achieving maintainable profits of approximately £84 000. Which investment is likely to give the better return on capital investment?

Own plant

Investment required	£500 000
Estimated maintainable profits	£ 75 000
Return on capital employed	$\dfrac{£\ 75\ 000}{£500\ 000} \times 100 = 15\%$

Existing company

Likely takeover price	£600 000
Estimated maintainable profits	£ 84 000
Return on capital employed	$\dfrac{£\ 84\ 000}{£600\ 000} \times 100 = 14\%$

On this basis it would appear that establishing its own plant could yield a marginally greater return.

17.5.4 Turnover of stock and debtors

With the efficient utilization of cash resources being such an important feature of a successful business, there are two common ratios which endeavour to measure the efficient use of that resource. Earlier in this chapter we said that the purchase of stock or the granting of credit to our customers represented an outflow of value from the business. Unnecessary cash resources tied up in stock and debtors is therefore extravagant and the quicker these assets can be turned over and realized as cash, the better. They will, of course, always require renewing but provided the cycle is maintained the resources used are contained at a minimum but commercially acceptable level. Difficulty can arise as a result of changes in the level of business activity. This is mainly reflected in the figure of sales, since to support a higher level of sales most likely requires a higher level of stocks and credit being given to customers. However, it is important to identify that despite the increase (or decrease) in activity the efficient use of cash resources is not being eroded.

Problems of comparison will arise when turnover is seasonal because it is probable that stock will be built up throughout one period in order to cater for the seasonal demand in another period. Additionally, debtors will probably be greater during the period of high seasonal sales, than for the rest of the year.

These performance measures are usually expressed in days when applied to debtors and when applied to stock as the number of times stock is turned over in a year.

Illustration

Assuming a level of turnover constant throughout the year, we can assess the measure of stockholding and days of debt in the company as follows:

Turnover in the year	£1m
Stock	£200 000
Debtors	£141 781

Stock

Stockholding is turned over: $\dfrac{£1\ 000\ 000}{£200\ 000} = 5$ times in the year

Debtors:

Debtors of £141 781 includes VAT at 15 per cent. It is therefore necessary to adjust either debtors or turnover to compensate for this in the calculation. Therefore debtors represent:

$$\frac{£141\ 781}{£1\ 000\ 000 + 150\ 000} \times 365 = 45\ \text{days}$$

17.5.5 Gearing

Some of the most spectacular business failures in recent years have arisen in companies which have been said to be highly geared. Gearing is the ratio of borrowed capital from sources such as banks and other investment institutions to the value of the equity in the business owned by the proprietors. The higher the ratio the more vulnerable the business is to problems of liquidity arising from higher interest rates or a downturn in the company's performance, thereby giving difficulties in repaying the loan debt. A business can become highly geared through rapid expansion based upon borrowed loan capital, as opposed to funding from proprietors or profits.

Illustration

Company A and Company B both have share capital and reserves (otherwise known as shareholders' funds) amounting to £100 000. Company A has bank loans amounting to £20 000 and Company B has bank loans of £50 000. Gearing can be calculated as follows:

	Company A	*Company B*
Shareholders' funds	£100 000	£100 000
Bank loans	£ 20 000	£ 50 000
Gearing	$\dfrac{£\ 20\ 000}{£100\ 000} = 20\%$	$\dfrac{£\ 50\ 000}{£100\ 000} = 50\%$

17.6 Summary

In this chapter we have described in outline the principles which lie behind financial and management accounting, costing and methods of evaluating business performance. It is important that systems analysts gain a good understanding of the financial management of the enterprise for which they are developing new systems. The material in this chapter should help in understanding management's role in the control of business operations and consequently enable systems analysts to develop appropriate new systems.

Appendix Forms Design

Introduction

Forms are an important means of communication. They are not cheap. A medium sized commercial organization probably uses several hundred different types of form—many of them in considerable volume. Their production is costly—but this is grossly outweighed by the cost of entering information on the forms and its subsequent use. Therefore it pays to exercise care in the design of a form to ensure that it is easy to complete and use.

Before introducing a new form its need must be firmly established. It may be that the transfer of information can be achieved in other ways using forms already in existence. A new form must be justified by assessing the following—is it the most efficient means of (1) obtaining information, (2) disseminating information, (3) storing information?

It is desirable to have a forms controller within an organization to whom all proposals for new forms must be submitted along with the case for their justification. Good forms control can result in (1) elimination of unnecessary forms, (2) periodic reviews of necessity, (3) reduction of production cost by control over paper and printing purchases, (4) minimizing stock holding, (5) efficient form completion and utilization.

Aids to design

The tools of the trade are simple—pencils, rulers in both metric and imperial measurements, a printer's typescale, scissors, eraser, obliterating fluid, glue. If a lot of form design work is done a drawing board is desirable. Ordinary drawing paper may be adequate for occasional form design but for more frequent form design it is worth buying guide sheets such as can be purchased from large stationers.

Methods of printing

The choice of method of printing depends on many factors: (1) number of copies required; (2) the use to which the form will be put; (3) the required quality and appearance; (4) time available for production; (5) cost of production; (6) the printing facilities available within the organization.

The main methods are as follows:

(1) *Letterpress printing* This is a professional task performed by specialist organizations. The printing surface may consist of metal types (formes) set on a flat bed or of curved metal, plastic or rubber plates fitted to cylinders. The paper feed may be either single sheets from a stack or hopper or continuous from a reel (as with what is called a web-fed rotary machine). As a very rough guide, letterpress would not be considered economical for runs of less than 10 000 copies.

(2) *Offset lithography* This differs from letterpress in that it uses a non-raised printing surface and relies upon the fact that grease and water do not mix. The parts of the printing plate which are to be printed are made grease receptive, the remainder is dampened. Inking rollers are impregnated with a grease-based ink so only the grease-receptive areas of the printing plate will receive ink. The inked form image is then 'offset' on to a rubber cylinder and thence on to paper.

(3) *Offset duplicating* This employs the same principle as that described above for offset lithography presses. The plates may be made of paper, plastic or metal and the image may be produced indirectly by photographic means or directly inscribed by pen, pencil or typewriter. Paper plates will give from 50 to 2000 copies; plastic plates up to 5000, but in some circumstances may yield much more; aluminium plates up to 25 000; zinc plates (which give best results) up to 50 000 copies.

(4) *Dyeline, electrostatic and spirit duplicators* These are suitable only for very small quantities of forms where appearance and quality are not important.

Paper

Most organizations will make use of paper sizes based on the International Standards Organization (ISO) as set out in British

Designation	Size (mm)
A0	841 × 1189
A1	594 × 841
A2	420 × 594
A3	297 × 420
A4	210 × 297
A5	148 × 210
A6	105 × 148
A7	74 × 105
A8	53 × 74
A9	37 × 53
A10	27 × 37

Standard 4000. The commonly used A range of paper sizes is shown above.

Other factors affecting choice of paper are as follows:

(1) *Surface quality* A slightly matt surface aids form completion by pencil or ballpoint pen; a more polished surface is needed for pen and ink; a smooth surface aids handling. Papers with a rough surface tend to cling together.

(2) *Erasing characteristics* Some paper will provide a good writing surface after erasure; on the other hand, for security reasons, a paper quality may be needed which makes erasure difficult.

(3) *Handling and storing requirements* Here, thickness, bulk, weight and toughness have to be considered. The weight of paper is measured in grams per square metre (gsm); the most common weights are 45 gsm (e.g. typing copy paper), 70 gsm (e.g. single-sided forms not subject to repeated handling), 80 gsm (e.g. double-sided forms with repeated handling), 100 gsm (e.g. for OCR devices).

(4) *Colour* Coloured paper may be needed to distinguish different documents and facilitate sorting and identification. However, there are some things to bear in mind: some people are colour blind; some colours cause eye-strain; some colours fade; some colours are not suitable for certain types of photocopying; coloured paper is slightly more expensive than white; the use of more than one colour of printing on a page increases the cost; printing in pale colours tends to be illegible, especially for people with colour blindness—stronger, mixed colours such as brown, greyish blue or green are better; red and green should not normally be printed side by side as it impairs legibility and creates disturbing visual effects; order of legibility is given by Le Courier's table on page 367.

Order of Legibility	Printing Colour	Background Colour
1	Black	Yellow
2	Green	White
3	Red	White
4	Blue	White
5	White	Blue
6	Black	White
7	Yellow	Black
8	White	Red
9	White	Green
10	White	Black
11	Red	Yellow
12	Green	Red
13	Red	Green

Design stages

The stages of design may be summarized as follows (following approval to go ahead by the forms controller):

(1) Define objective of form.

(2) Specify its data content (the NCC Clerical Document Specification form may be used for this purpose).

(3) Decide upon quantity required and likely method of production.

(4) Determine paper size.

(5) Using a guide sheet: enter title; enter form reference number; indicate position of any specially located material, e.g. address panel for a window envelope; within space left for main contents, enter the items, notes and instructions.

(6) Review with the user; revise using new guide sheet, if necessary.

(7) Produce fair specimen.

(8) Submit to forms controller for approval.

(9) Consider limited production run and field test.

(10) Revise as necessary.

(11) Order form (via forms controller).

Content and layout

Appearances are important. Forms which look complicated and

time-consuming to complete, or are unattractive to the eye, are not likely to do the job they are intended to do. Words and phrases which may be familiar to the designer may be misunderstood by the user—so explanatory notes may be necessary. Avoid brusqueness and officious-sounding words—a little politeness does not come amiss! It is much better to say, 'Please complete part B...', instead of, 'You must complete part B...'.

The form has to satisfy the needs of the person completing it and the subsequent user. Compromises have to be made to resolve the conflict of interest. This frequently arises with computer-system forms. A form carrying input data has to be designed not only to ensure that accurate data is recorded at source but also to facilitate data entry (say by a keyboard operator). A form produced by a computer line-printer has to be designed not only to convey information in a clear and unambiguous way to the recipient but also to facilitate its production on the line-printer and other ancillary equipment. Increasingly the user's interest is being regarded as paramount—it is useless efficiently processing incorrect input data or efficiently producing confusing output information.

The entries on a form should provide a logical sequence—bringing together related material. The flow should be left to right from top to bottom. Grouping of information by the use of panels or blocks is helpful.

Explanatory notes should be kept to a minimum, otherwise they will clutter up the form and confuse the user. All notes which need to be read before a form is completed should be placed at the top of the form. A note relating to a particular item should be placed in conjunction with the item.

Printed lines are known as 'rules' in the trade. They should be used sparingly. Dotted lines may be used to guide the insertion of manuscript entries; they should not be used where a typewriter is to be used as they may create a registration problem and add to the cost of completion. For typewriter completion, entries are best made in columns so that the typist can use tabulation stops.

Stippling or shading is effective for highlighting parts of a form—but it may add to the cost of printing.

The form title should be as simple as possible—and the word 'form' should be avoided if possible (instead use note, report, list, analysis, application, or some such meaningful word).

Care has to be taken over the allocation of space to headings describing data or information entries—see Figure A.1. A balance has to be struck between space needed by a heading and that needed for the actual entry; the amount of space should be determined by the entry requirements, not by the heading—*see* Figure A.2.

There are a number of ways in which choices can be dealt with. Here are some examples:

Colour preference (ring the one you like most)

Black   Blue   Red   (Green)   Brown   Yellow

Colour preference (insert × in the appropriate box)

Black	Blue	Red	Green	Brown	Yellow
☐	☐	☐	☒	☐	☐

COLOUR PREFERENCE

Put a tick against the one you like most

Black
Black
Red
Green √
Brown
Yellow

Care must be taken to ensure that the captions are unambiguously related to the boxes or spaces provided for entries. A bad example is

Colour preference (place a cross against the one you like most)

Black   Blue   Red   Green ×   Brown   Yellow

A better example is

Colour preference (place a cross in the box for the one you like most)

Black	Blue	Red	Green	Brown	Yellow
☐	☐	☐	☒	☐	☐

It is preferable to stick to one method of indicating choice, otherwise the form appears confusing, like the following:

Insert a cross if you are seeking special discount ☐
We do/do not* purchase widgets from another supplier
*Delete as necessary
What is the name of your agent? ....
Circle the number of retail outlets you have
    1   2   3   4   5–20   more than 20

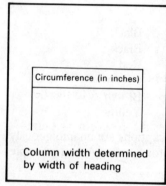

Fig. A.1

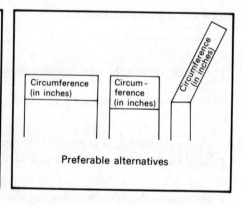

Fig. A.2

Forms destined to be input to a computer via a keyboard operator require special attention. In addition to satisfying the needs of the originator they should also satisfy the following:

(1) They should be clear and easy to read.

(2) The data to be keyed should be distinguishable from other matter.

(3) They should be easy to handle.

(4) They should facilitate smooth working rhythm (e.g. all data on one page of a sheet to avoid turning over).

(5) There should be a clear relationship between the data and the input medium.

(6) They should have a consistent format.

Keying efficiency is generally improved if items of data are presented vertically rather than horizontally. If items are staggered in regular groups, accuracy is improved, e.g. as in Figure A.3.

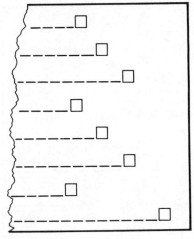

Fig. A.3

Documents which are intended for input to a computer via OMR, OCR or MICR devices must be designed to conform to the device specifications.

Output forms should be designed with the recipient in mind— indeed if the recipient is within the organization he should be closely involved in the design. The constraints of the output device will obviously limit the scope. The paper for line-printers may be (1) single part; (2) multi-part with carbon interleaves; (3) self-copying sets of paper (NCR); (4) special—for diazo or offset duplicating; (5) plain or preprinted.

Summary

Form design is a specialist activity. It is important because it often represents the interface between the user and the system. A badly designed form costs a lot of money—throughout the life of the system. It is desirable to have a focal point within an organization to control the authorization and production of forms.

Fig A.6

Equipment which are intended for input to a computer via OMR, OCR or ICR devices must be designed to conform to the device specifications.

Output items would be designed after the confirmation of this user-led throughput for whom the organization. It should be closely involved in the design. The containers or the output device will obviously limit the scope. The paper (or component) may be (1) single part (2) multi-part with carbon influences (3) self-carbon sets or paper (NCR) with duplication for each application (5) plate or perforation.

Summary

Form design is a particular activity of a important nature. Problem represents the interface between the user and the system. Poorly designed forms pose a lot of problems throughout the life of the system. It is thus useful to use a good form design procedure to minimize any occurrence and consequences of this.

Index

Index